PLANT ECOLOGY

Recent Advances in Environmental Ecology Series

PLANT ECOLOGY

A. P. Diwan
D. K. Arora

ANMOL PUBLICATIONS PVT. LTD.
NEW DELHI - 110 002 (INDIA)

ANMOL PUBLICATIONS PVT. LTD.
Regd. Office: 4360/4, Ansari Road, Daryaganj,
New Delhi-110002 (India)
Tel.: 23278000, 23261597, 23286875, 23255577
Fax: 91-11-23280289
Email: anmolpub@gmail.com
Visit us at: www.anmolpublications.com

Branch Office: No. 1015, Ist Main Road, BSK IIIrd Stage
IIIrd Phase, IIIrd Block, Bangalore-560 085 (India)
Tel.: 080-41723429 • Fax: 080-26723604
Email: anmolpublicationsbangalore@gmail.com

Plant Ecology

Reprint, 2010
ISBN 81-7488-121-2

PRINTED IN INDIA

Printed at Mehra Offset Press, Delhi.

Preface

Ecology is the most active and most fascinating subject among the biological sciences. It seems that ecology is still in the formative stage and offers an alluring front for the advancement of knowledge of science. Most field studies of living organisms now have an ecological orientation. Ecology is the special term for the environment biology and the study of ecology is the attempt to understand the relationship of plants and animals to their environments—where they live, how they live there and hopefully, why they live there.

This book consists of eighteen chapters on various aspects of plant ecology. The authors have collected all the relevant informations from different possible sources to make the book comprehensive as well as uptodate.

Any criticism and suggestion for improvement would be thankfully acknowledged.

AUTHORS

Preface

Ecology is the most active and [illegible] fascinating subject among the biological sciences. [illegible] that emerged [illegible] within the formative stage and offers [illegible] for the advancement of knowledge of [illegible] investigations now have [illegible] ecological [illegible] concern for the environment [illegible] is the attempt to understand [illegible] of plants [illegible] to their environments [illegible] and [illegible] why they [illegible] here.

The book consists [illegible] of plant ecology. [illegible] all [illegible] relevant information [illegible] the [illegible] comprehension [illegible]

Any [illegible] acknowledged [illegible] would be thankfully acknowledged.

Authors

Contents

1

Introduction

At present ecology is among the most active and most fascinating of the biological sciences. It seems that a really general ecology is still in the formative stage and offers an alluring front for the advancement of knowledge. Most field studies of living organisms now have an ecological orientation. Basic attitudes and principles of ecology have also permeated most of the biological sciences and some other sciences. Ecology is the special term for the environment biology.

The word ecology (Oekologie) is derived from the Greek word, "olkos" (meaning house, abode, dwelling) and logos (meaning study or discourse). Thus literally, ecology is the study of "houses" or in a broader sense "environments". As it is concerned especially with the biology of groups of organisms and functional processes on the land, in the ocean as well as in fresh water, one may therefore define ecology as the study of structure and function of nature, including the living world. Thus ecology is the attempt to understand the relationship of plants and animals to their environments—where they live, how they live there and hopefully, why they live there.

According to Haeckel (1869) "ecology is the science treating of reciprocal relations of organisms and the external world."

Elton (1927) called ecology "scientific natural history". According to Taylor (1936) "ecology is the science of all the relations of all organisms to all their environment." Abercrombie *et al.*, (1954) opinioned that the study of the relations of animals and plants, particularly of animal and plant communities, to their surroundings, both animate and inanimate is called ecology. Odum (1963) defined ecology as "study of the structure and function of nature." Hughes and Walker (1965) defined ecology as "study of the relationships between organisms and their environment." Ecology is that branch of science which deals with the relationship between living things and their physical environment together with all the other living organisms within it (Phillipson, 1970).

General ecology deals with the biota (flora and fauna) and its environment. Ecology, as used to-day is a science which states organisms in relation to their environment: a philosophy in which the world of life is described in terms of natural processes; an art requiring skill and having a plan and a pattern within which many activities may be centred. The man emphasis of ecology is on the relationship between organisms and groups of organisms and their external environment.

HISTORY OF ECOLOGY

The history of ecology can be traced back to prehistoric man who took into account the environmental factors for hunting, trapping animals, finding edible vegetation and finding out of shelter to protect himself from hardships of nature. When man understood the importance of environment, it took a religious turn and man started worshiping sun, air, rains etc. The early Greek scientists and philosopheres also understood the importance of weather. First of all Hippocrates published a paper entitted "On Air, Waters and Places". After Hippocrates, Aristotle studied the habits of animals and environmental conditions. Later Theophrastus, a student of Aristotle wrote about plants communities and the types of plants found in different areas. Thus Theophrastus is considered to be first ecologist.

Study of natural sciences was resumed by Reaumur (1683-1757) after centuries gap. He published about six volumes on the Natural History of Insects which contained ecological information about insects. Baron Alexandert von Humbolt (1804) published about 26 volumes based on the data collected by him after exploring tropical and temperate South America. This gave an impaetus to other naturalists to study flora and fauna of South America. Henry W-Bates published work on termites, warrior ants and parasol ants in relation to environment. Richard Spruce explored Amazon and Negro rivers while Alcide d' Orbigny (1826) collected specimens in Andes Mountains and Bolivia. Edward Forbes studied the flora and fauna of Mediterranean Sea. Joseph Hooker studied the flora and fauna of polar continent. Louis Agassiz (1846-1873) published "Contribution to the Natural History of the United States." He first founded Marine Laboratory in United States.

Darwin published "Journal of Researches into the Natural History and Geology of the Countries visited during the voyage of H.M.S. Beagle." Edward Forbes (1846) published a paper on palaeo-ecology of the British Isles. Alfred R. Wallace published three books:

(i) The Malaya Archipelago,

(ii) Island life and

(iii) The Geographical Distribution of animals.

Daven port (1903) published a paper on "The animal ecology of Cold Spring Harbor." S. A. Forbes (1907) described the distribution of Illinois fishes. E. Warming (1909) established the interdependence and close relationships between plant and animals. V.E. Shelford (1907, 1908) worked on Tiger Beetle. Allee, *et al.*, (1949) published "Principles of Animal Ecology." In 1954 Andrewartha and Birch published scholarly work "The Distribution and Abundance of Animals."

Origin of Plant Ecology

While it is true that ecology as a science arose near the close of the last century, but there have always been a few naturalists interested to some degree in environmental relations. The philosopher, Theophrastus (370—285 B.C.) may well be regarded as the first ecologist in history because he wrote and quite sensibly too, of the communities in which plants are associated, the relation of plants to each other and to their physical environment. He also studied tne features of water plants (swampy and marshy plants) and plants of dry and acid plants.

There was an era of plant geography during which general studies of vegetation and geographical distribution of individual plants were made in eighteen century. The works of Humboldt Schouw, A, de Candolle are the characteristics of this era. In the period 1838—1895 the study of plant formations was made. A plant formation was recognized as fundamental unit of vegetation. This period was the beginning of the study of plant succession. The term ethology was proposed by the French zoologist named Hilaire in 1859 to designate the subject. The term ecology was, however, proposed in 1885 by Reiter, a zoologist. In 1886, Haeckel also another zoologist defined ecology as the study of the reciprocal relations between organisms and their environment.

The first general work devoted to ecology was by the Danish biologist J.E.B. Warming in 1895 followed by Swiss biologist A.F.W. Schimper in 1898. Warming published his book Oecology of plants and Schimper published a book named Plant geography upon a physiological basis. The period, 1895—1916 is the early part of modern plant ecology. The contributions of Schimper, Drude and Warming are characteristics of this period. In this period the study of habitat along with the exact determination of physical factors was made. The plant ecology was divided into autecology and synecology. After 1916, ecological research was greatly expanded by the establishment of ecological societies and publication of ecological journals. After world war II, the United Natións Educational Scientific and Cultural Organisation (UNESCO) started desert and arid zone studies. Numerous regional ecological stations have been set up.

Indian Work Ecology

In India Ecological work has been started in the begining of the 20th century. In this century names of Prof. Dugdeon, Prof, Barucha, Prof. R. Misra may be mentioned as pioneer worker of ecology. "Indian forest Ecology" was published by Dr. G.S. Puri, Prof. R Misra, L.P. Mall, D. K. Tiwari etc., conducted Autecological studies of a number of plants. Prof. Misra, Pandeya and K.P. Singh contributed much to the Production Ecology. Still the work on Grass land Ecosystem is being done by Dr. K.C. Misra, J.S. Singh, L.P. Mall etc. Important Ecological Research centers in India are Banaras Hindu University, Saurastra Univ., Vikram Univ., Kurukshetra Univ., Punjab Univ. etc.

The environmental biology is called as Ecology which is a field of special interest these days. The term ecology, first of all described by Ernst Haeckel (1866), a German Zoologist, is derived from Greek word "Oikos" meaning "hose". Geoffroy St. Hilaire (1859) a French Zoologist proposed the term "Ethology" instead of ecology for the study of living being in relation to their environment. Lankester in 1889 used the term "Bionomics" for the study of organisms and their relationship with the environment. The ecology is the study of biological interrelationship between organisms and their environment. In other words the ecology is concerned with the biology of groups of organisms and their relationship with the environment such as land, sea, fresh water etc. The ecology is the combination of biochemistry, biophysics, morphology and physiology. Modern ecologists define ecology as "the study of the structure and function of nature." Here by nature we mean the animals and plants or organisms. An organism may be defined as "a self regulating and self perpetuating physico-chemical entity which is in state of perfect balance with the environment." The term environment includes the surroundings of the organism which has a direct influence on it.

SUBDIVISIONS OF ECOLOGY

Early ecologists have recognized two major subdivisions of ecology in particular reference to animals or to plants, hence animal ecology and plant ecology. But when it was found that in the eco-systems plants and animals are very closely associated and inter-related then, both of these major ecological subdivisions became vague. However, when animals and plants are given equal emphasis, the term bioecology is used. Further, ecology is often broadly divided into autecology and synecology. Autecology deals with the ecological study of one species of organism. Thus, an autecologist may study the life history, population dynamics, behaviour, home range and so on, of a single species, such as the Mexican free-tailed bat, Indian bull frog, or maize-borer, Chilo partellus, Synecology deals with the ecological studies of communities or entire ecosystems. Thus, a synecologist might study deserts, or caves or tropical forests. He is interested in describing the overall energy and material flow through the system rather than in concentrating on the finer details of a particular organism. In the words of Herreid II (1977) "the two types of study, autecology and synecology, inter-relate, the synecolosist painting with a broad brush the outline of the picture and autecologist stroking in the finer details."

Besides these major ecological subdivisions, there are following specialized branches of ecology:

1. Habitat ecology. It deals with ecological study of different habitats on planet earth and their effects on the organisms living there. According to the kind of habitat, ecology is subdivided into marine ecology (oceanography), estuarine ecology, fresh water ecology (limnology), and terrestrial ecology. The terrestrial ecology in its turn is classified into forest ecology, cropland ecology, grassland ecology, desert ecology, etc., according to the kinds of study of its different habitats.

2. Palaeoecology. It is the study of environmental conditons, and life of the past ages, to which palynology, palaeontology, and radioactive dating methods have made significant contribution.

3. Systems ecology. It is the modern branch of ecology which is particularly concerned with the analysis and understanding of the function and structure of ecosystem by the use of applied mathematics, such as advanced statistical techniques, mathematical models, characteristics of computer sciences.

4. Community ecology. It deals with the study of the local distribution of animals in various habitats, the recognition and composition of community units, and succession.

5. Taxonomic ecology. It is concerned with the ecology of different taxonomic groups of living organisms and eventually includes following divisions of ecology: microbial ecology, mammalian ecology, avian ecology, insect ecology, parasitology, human ecology and so on.

6. Population ecology (Demecology). It deals with the study of the manner of growth, structure and regulation of population of organisms.

7. Applied ecology. It deals with the application of ecological concepts to human needs and thus, it includes following applications of ecology: wild-life management, range management, forestry, conservation, insect control, epidemiology, animal husbandry, aquaculture, agriculture, horticulture, land use and pollution ecology.

8. Evolutionary ecology. It deals with the problems of niche segregation and speciation.

9. Production ecology. It deals with the gross and net production of different ecosystems like fresh water, sea water agriculture, horticulture, etc., and tries to do

proper management of these eco-systems so that maximum yield can be get from them.

10. Human ecology. It involves population ecology or man and man's relation to the environment, especially man's effects on the biosphere and the implication of these effects for man.

11. Geographic ecology (ecogeography). It concentrates on the study of geographical distribution of animals (zoogeography) and plants (phytogeography), and also of palaeoecology and biomes.

12. Radiation ecology. It deals with the study of gross effects of radiations and radioactive substances over the environment and living organisms.

13. Ecosystem dynamics. It deals with the ecological study of the processes of soil formation, nutrient cycling, energy flow, and productivity.

14. Ecological genetics. An ecologist recognised kind of genetic plasticity in the case of every organism. In any environment only those organisms that are favoured by the environment can survive. Thus, genecology deals with the study of variations of species based upon their genetic potentialities.

15. Ecological energetics. It deals with energy conservation and its flow in the organisms within the ecosystem. In it thermodynamics has its significant contribution.

16. Chemical ecology. It concerns with the adaptations of animals of preferences of particular organism like insects to particular chemical substances.

17. Pedology. It is a branch of terrestrial ecology and it deals with the study of soils, in particular their acidity, alkalinity, humus contents, mineral contents, soil-types, etc., and their influence on the organisms.

18. Physiological ecology (ecophysiology). The factors of environment have a direct bearing on the functional aspects of organisms. The ecophysiology deals with the

survival of populations as a result of functional adjustments of organisms with different ecological conditions.

19. Sociology. It is the study of ecology and ethology of mankind.

20. Ethology. It is the interpretation of animal behaviour under natural conditions. In it, often, detailed life history studies of particular species are amassed

SCOPE OF ECOLOGY

Ecology is a multidisciplinary science and it includes not only the life sciences but chemistry, physics, geology, geography meteorology, climatology, hydrology, palaeontology, archeology, anthropology, sociology and mathematics and statistics as well. For explaining the behaviour of an organism or biotic community in a given environment, an ecologist has to integrate the data which is obtained from many sources—morphology, taxonomy, genetics, physiology, soil science, climatology, geology, physics and chemistry.

The scope of ecology is quite vast. The study of ecological principles provides a background for understanding the fundamental relationships of the natural community and also the sciences dealing with particular environment such as forest, soil, ocean and inland waters. Many practical applications of this subject are found in agriculture, horticulture, forestery, limnology oceanography, fishery biology, biological survey, game management, pest control, public health, toxicology, pollution control, conservation, etc. Ecological knowledge helps in discovering new sources of food (algae, krill, etc.), new unpolluting sources of energy (*i.e.*, solar energy), and new methods of pest control such as biological control which causes no environmental pollution. By applying certain ecological techniques, some ecologists have been successful in diagnosing the cause of desertness of certain Australian deserts and they investigated that these deserts lack certain trace elements like zinc, copper and molybdenum, so they called them trace element deserts (Anderson and Underwood, 1959). Now they have cured their ecological diseases and have converted them into new rich agricultural lands.

Approach to the Study of Ecology

The ecology has several methods of approaches which are given below:

(i) Zoological approach. In Zoological approach mainly the animals are considered while the habitat and plants are considered as the part of the environment.

(ii) Botanical approach. In Botanical approach mainly the plants are considered while the animals and the habitat are considered secondarily.

(iii) Biotic approach: Includes the flora and fauna of a particular region. This approach considers the inter-relationship between plants and animals, between the plants themselves, between animals themselves and between living being and environment.

(iv) Habitat approach. When the study is focused mainly on the habitat, such as freshwater, marine, estuarine habitat and terrestial habitat etc. It includes forest ecology, grass land ecology, cropland ecology, desert ecology, fresh water ecology, marine ecology etc.

(v) Productivity approach. It considers the yield resulting from the energy relationships between the organisms and habitat.

(vi) Population approach. It includes group relations and deals with population dynamics in pure or mixed strands.

(vii) Ecosystem approach. This approach is not specific instead it includes habitat, plants and animals together *i.e.*, an ecosystem, like a pond or lake or part of the ground.

(viii) Gene-ecology. It is concerned with the genetic make up of a species or population in relation to environment.

(ix) Conservation ecology. It deals with the proper management of natural resources for the benefit of human beings.

(x) Ecological Energetics. It is a recent branch of ecology which deals with the energy conservation and flow in the organisms within the ecosystem.

SCOPE OF ECOLOGY

The ecologist focuses his attention primarily on the inter-relationship between the organism and its environment. The possible scope of ecology becomes wide the due to variable environmental conditions and more abundance of plants and animals. The main function of ecology, however, is to show the general principle under which the natural community and its component parts operate. Then these may be applied to the interpretation of activities of the particular plants and animals found in a given situation.

The biota (flora and fauna) of an area may be identified and counted. The physical forces may be recognised at work, yet, neither an account of biota, nor a description of the habitat makes an ecological investigation. Similarly, if one makes a list of hydrophytes or mesophytes or xerophytes without any consideration of the relation of the occurrence of these species to other environmental factors, he cannot be called an ecologist. Modern ecology is concerned with the functional interdependence (relationship) between living things and their surroundings and it is a field subject.

The study of ecological principles provides a background for further detailed investigation not only into the fundamental relationship of the natural community but also into sciences dealing with particular environments such as forest, soil, ocean and inland waters. Many practical applications of this subject are found in agriculture, biological survey, game management, pest control, forestry and fishery biology. An understanding of ecology is more valuable for suitable conservation, whether in relation to soil, forest, wildlife. water supply or fishery resources. In a wider sense ecology is also significant for us as citizens because man himself is most important element in the enviro.ment. Thus the scope of ecology is not narrow but wide.

Factors of the Environment

The environment is of two types physical or non-living or abiotic and living or biotic environment. The abiotic or non-living environment includes medium in which an organism live and climate which surrounds it. For living organisms, plants as well as animals, four types of media are available soil, water, air and bodies of other organisms.

Soil is the upper most strata of the earth which contains organic matter which is capable of nourishing vegetation. Thus the chemical composition of soil is important in determining the presence of plants and animals. The texture of the soil has also its own significance because it regulates the moisture conservation.

Water forms the important medium for aquatic animals and plants. Water plays an important role in respiration more so because protoplasm is composed of large quantity of water besides other constituents. It also helps the body in metabolism.

Air is present all around the earth in the gaseous form. Although it is a mixture of gases yet oxygen and nitrogen are present in 21 parts and 78 parts respectively. The remaining one part contains carbon-dioxide, ozon, neon, argon etc. Other constituents like organic matter, dust, micro-organisms, salt, sulphates, water vapour etc., are also present. The percentage of these substances vary very much depending upon time and place and has considerable effect on climate, weather, radiation and comfort.

Bodies of organisms are found in other animals and they live as parasites. The parasites fullfil their requirements of oxygen from internal environment of the host and protect themselves from digestive or other juices. There are very little fluctuations in such environment.

The biotic or living environment may be explained by two popular terms the intra-specific and inter-specific. The intra-specific relationship denotes the relationship between the members of the same species while inter-specific-relationship includes the relationship between the members of different species. The intra-specific relationship is governed by reproduction, assistance, competition and definite hostitity.

The reproduction affects the intra-specific-relationship. During sexual or asexual reproduction number of individuals are increased and if the rate of reproduction is high it will result to over production and leads to competition or struggle for food and space.

The assistance means the parental care or protection given by the parents to their helpless young ones.

The competition results from over production of offsprings and limited food and shelter.

The definite hostility involves struggle over territoral limits and in the selections of mates.

According to Darwin's theory also whenever there is rapid reproduction and limited supply of food there is always struggle for exhistance and then only the fittest survives or the survival of the fittest.

The inter-specific-relationship is governed by competition for food, prey predator relations, host-parasite relations, commensalism symbiosis and slavery.

The competition amongst the members of the same species is very fierce because the needs of the members of the same species are the same as regards to food, shelter, selection of mate etc.

The prey-predator relationship denotes the feeding of one animal over other. The animal which feeds is the predator and which is eaten, is called as the prey. The predators can survive only when the prey are available or in other words the prey cannot survive if the predators are many in number.

The host-parasite relation may be complete or in part parasite may depend on its host partly or completely.

The commensalism is the phenomenon in which two individuals lie in mutual reation with each other but none is dependent on other.

The symbiosis is a close physiological relationship that exists between members of different species completely dependent on each other.

The slavery includes making of slaves. When certain social animals capture the other animals, make them slaves and utilize their services, the phenomenon is called as slavery.

REVISION QUESTIONS

1. Trace the origin of the term 'Ecology', in history?
2. How will you define ecology?
3. What is the scope and objectives of ecology?
4. Describe different subdivisions of ecology.
5. Describe, why ecology is significant for human being?

2

Weathering and Soils

Over 3 billion years ago, rain and snow fell upon the earth much as today. Solid rocks were reduced to grains of sand by the action of water, ice, and chemically active solutions. We know this from clues left in ancient rocks whose great antiquity has been established by radioactive dating. The processes by which rocks are broken into smaller particles and chemically decomposed have continued unabated down through the eras of geologic time. We term these most enduring of geologic processes weathering. Weathering is of enormous importance to all of us. It causes solid rock to crumble and thereby facilitates erosion. Weathering provides much of the sediment for the making of sedimentary rocks. Without weathering there would be no soil for the growth of plants that are vital to ourselves and all other forms of life.

Weathering is a phenomenon that combines two categories of processes. On the one hand are those processes that result in the fragmentation of once solid rock without chemical change, as when water freezes and expands in crevices so as to cause the rock to break apart, or when rocks are wedged apart by the growth of roots from trees and shrubs. These processes of physical disintegration are designated mechanical weathering. In contrast, chemical weathering involves the chemical decomposition of rocks as their minerals react with water and air. Mechanical and chemical weathering work together in the destruction of rocks at the earth's surface, although one may predominate over the other. In areas of extreme cold or aridity, for example, mechanical weathering may predominate, but in warm wet regions, chemical decomposition is the predominant kind of weathering.

Although mechanical breakup and chemical decay of a rock are partners in weathering, it is advantageous to discuss these two components of weathering separately, because disintegration involves largely mechanical processes, whereas decomposition entails the chemical interaction of water, gases, and minerals.

MECHANICAL WEATHERING

The chief agents of mechanical weathering, or disintegration, are frost action, temperature changes, unloading, crystal growth, and the wedging action of plant roots. Many different factors influence the efficiency of these agents in causing disintegration. The composition and texture of the rocks being weathered and the presence of joints, fractures, and voids clearly affect the rate at which solid rock can be reduced to rubble. Mechanical weathering is also affected by climate, topography, and the length of time over which weathering agents have been operating. In general mechanical weathering predominates over chemical weathering in regions characterized by extreme cold or in warm arid regions. By definition, mechanical weathering involves the physical disintegration of solid masses of rock into loose fragments. There is little chemical change in the rock itself. Thus, a chemical analysis of the disintegrated rock would be similar to that of the parent rock.

Frost Action

Water expands by about 9 per cent when it freezes. If tha water freezes in a confined space, pressure caused by the expansion may cause rock masses to be pushed apart and ruptured. Such frost action has an important effect in mechanical weathering in that the freezing water is capable of exerting thousands of pounds of pressure per square inch. Many of us have become aware of the effects of frost action during severe winters when concrete roads and side-walks are cracked by alternate freezing and thawing. The way freezing water disrupts solid rock is similar to the manner in which it breaks up streets and side-walks. After water has filled a crack in the rock, the water at the lip of the crack may freeze Once this has occurred, the water deeper in the crack is sealed off, and as it begins to freeze and expand, it pushes against the walls of the fracture, and thereby widens the space between those walls. In a subsequent thaw, fragments of rock may slip down into the crack and act as wedges to hold it open. These processes are most prevalent in areas where water is abundant and where temperatures drop to below freezing at night and then warm during the day. Mountainous regions in temperate zones have this kind of daily temperature change, and in such rugged regions frost action is an important weathering process.

Insolation

Rocks, like many other solids, expand when they are warmed and contract when they are cooled. The heat is provided by the sun and so geologists refer to the weathering that may result as insolation weathering. It would seem logical that repeated day-time heating and nighttime cooling of rocks (with concurrent expansion and contraction) would weaken the boundaries between grains and eventually cause them to separate from the rock mass (Fig. 4.2). In laboratory experiments conducted by David T. Griggs at Harvard University, however, repeated heating and cooling did not cause fragmentation of the rock. Unfortunately, the experiments were unable to assess the amount of strain induced by millions of expansions and contractions over thousands of centuries. Most geologists believe that insolation is only a minor contribulor to mechanical weathering. Its effect is usually obscured by other weathering processes.

Perhaps in desert regions where there are large-scale fluctuations in temperature, insolation may have some effect in weathering rocks.

Unloading

Quarrymen and miners are well aware that when the heavy weight of rock is removed from a mine with consequent loss of support to the surrounding rock, the outside pressure may sometimes cause o veritable explosion of rock fragments into the excavations. In nature, erosion may similarly remove large volumes of surface rock, thus lightening the load on deeper rock and allowing it to expand. Because the mass of rock is still confined on all sides, it can only respond to the release of pressure by expanding upwards. As it expands it will rupture and form joints that are roughly parallel to the surface topography. The process is called unloading, and the joint systems are referred to as sheeting. Sheeting is often seen in granite and other massive crystalline rocks that have been laid bare by glaciation or running water. It can also be readily observed along the upper walls of quarries, where it may even facilitate the removal of the stone. Sheeting provides planar passageways along which solution and frost action may proceed vigorously. Half Dome at Yosemite National Park and Stone Mountain, Georgia, are examples of mountains whose shape has been controlled, at least partially, by sheeting.

Saline Crystal Growth

The growth of crystals, especially crystals of sodium chloride, calcium sulpaate, cr magnesium sulphate have been found to cause scaling in exposed rock surfaces and to dislodge grains and crystals from their parent rock. The process is particularly effective in porous rocks in which crystals exert large expansive stresses as they grow. Disintegration of building stone because of the growth of crystals in pore spaces has been a vexing problem for architects involved in the preservation of historically important buildings. The Temples at Luxor, Egypt, for example, have been seriously damaged by alternate solution and crystallization of salt. During dry spells, magnesium (and calcium) sulphate have been known to crystallize along zones of weakness in the dolomitic building stones of London's Houses of Parliament, thereby causing spalling and crumbling at a rate that has caused considerable anxiety among restoration specialists. The magnesium sulfate is formed as a result of a chemical reaction between coal smoke and the maguesium in the dolomite.

Root Wedging

When one observes the growth of a plant from a seed, it is apparent that even a frail seedling is capable of exerting sufficient force to push aside relatively firm soil. As they grow into rock fractures and expand, the roots of large plants such as trees can exert correspondingly greater forces and are capable of widening cracks and accelerating the rate of disintegration by significant amounts. When the plant dies and the roots decay, they leave openings in which freezing water may accumulate and further widen the void space. Plants also react with rocks chemically, as will be described in the following section.

CHEMICAL WEATHERING

Chemical weathering refers to the decomposition of rocks at or near the earth's surface and under relatively low temperatures. During chemical weathering, water and chemically active water solutions as well as oxygen and carbon dioxide from the atmospher attack the minerals comprising rocks. The susceptible parent minerals are frequently altered to softer secondary minerals, as when feldspars are converted to clay minerals and lose certain ions to solution. Chemical weathering is an exceedingly important process, for it leads to the formation of soils and in some parts of the world has resulted in the concentration of iron, aluminum, uranium, gold, and tin.

Relation of Disintegration to Decomposition

The processes of chemical weathering and mechanical

weathering are not truly independent of one another. Because of climatic factors, especially the amount and frequency of rainfall, one or the other process may be dominant in a particular location. For the most part, however decomposition and disintegration are interacting processes. For example, disintegration greatly promotes decomposition by reducing large intact rock masses to accumulations of smaller fragments. This fragmentation provides more total surface area for chemical attack than was originally present in the larger mass of rock. One can see what happens readily by imagining a block of rock measuring 2 meters on each edge. Such a block would have a total of 24 square meters of surface area. If the block is split once along each of three perpendicular planes, each resulting block would have 6 square meters of surface area or a total of 48 square meters for all eight pieces. Chemically reactive solutions are only able to decompose surfaces they can reach, and so the rate of chemichl attack is enhanced by increasing surface area. One can demonstrate this fact by applying a few drops of dilute hydrochloric acid first to an unfragmented piece of calcite and then to a similar piece of calcite that has been pulverized. Because of the greater amount of total surface area, the pulverized rock will display the more vigorous efferve scence.

Processes of Chemical Weathering

People sometimes are surprised to learn that many rocks exposed at the earth's surface are really not in chemical equilibrium with their environment but rather are unstable and are slowly but continuously reacting with atmospheric components to dissolve or change to new substances that are more nearly stable at the earth's surface. Geologically, weathering proceeds in accordance with a generalization called the "rule of stability," which states that a mineral approaches stability most closely in an environment similar to that in which it formed. Intrusive igneous rocks, of course, are formed under conditions of high temperature, high pressure, and a deficiency of free oxygen and fresh water. When such rocks are laid bare on the continents, they find themselves, in an environment of low temperature, low pressure, and abundant oxygen and water. Chemical change is inevitable, and the minerals most susceptible to change are those that formed under physical and chemical conditions most removed from conditions at the earth's surface. For example, amor g the common rock-forming silicate minerals, olivine forms at high temperatures and pressures early in the crystallization of a magma. Consequently, it rapidly weathers in the environments that exist at the earth's surface. Quartz forms much later under less extreme conditions of temperature and pressure and is less susceptible to weathering. In reference to the Browen Reaction Series it is apparent that minerals nearer the top of the series generally weather more rapidly than those near the base.

We noted how silicon-oxygen tetrahedra are joined together by eharing some oxygen ions and that the number of metallic ions needed to neutralize the crystal is reduced by this sharing. This increased oxygen sharing by silicon results in a greater number of strong covalentlike bonds between silicon and oxygen, and this in turn greatly increases the ability of the mineral to resist weathering. For example, the ratio of oxygen to silicon in olivine is 4, in pryoxene 3, in hornblende 2.7, in biotite 2.5 and in quartz 2. The diminishing ratios correlate nicely with greater resistance to weathering. All of the oxygen atoms are shared by silicon atoms in quartz, and this partially accounts for its great resistance to weathering.

The chemical weathering of rocks involves reactions that are interrelated, that may occur simultaneously, and that utilize water, oxygen, carbon dioxide, and organic acids. These processes include hydrolysis, oxidation, carbonation, solution, hydration, and chemical changes induced by the growth of plants.

Hydrolysis

Hydrolysis is a chemical reaction between a mineral and water. It involves a reaction between the H^+ or OH^- ions in the water and relatively active metallic ions such as sodium, calcium, potassium and magnesium. Hydrolysis is particularly important in causing the decomposition of silicate minerals. Although it may occur in the presence of pure water, in nature hydrolysis nearly always involves carbon dioxide. To illustrate, small quantities of carbon dioxide from the atmosphere or soil are dissolved in water to form carbonic acid.

$$\underset{\text{water}}{H_2O} + \underset{\text{carbon dioxide}}{CO_2} \leftrightarrows \underset{\text{carbonic aaid}}{H_2CO_3}$$

The carbonic acid may then ionize to form hydrogen ions and bicarbonate ions.

$$\underset{\text{carbonic acid}}{H_2CO_3} \leftrightarrows \underset{\text{hydrogen ion}}{H^+} + \underset{\text{bicarbonate ion}}{(HCO_3)^-}$$

Next, the hydrogen ions penetrate into the crystal lattices of silicate minerals such as potassium feldspar. They have little difficulty in gaining admission because of their small size, and once within the lattice, they disrupt the charge balance as they displace cations such as potassium.

$$\underset{\text{water}}{H_2O} + \underset{\text{hydrogen ions}}{2H^+} + \underset{\text{potassium feldspar}}{2(K\ Al\ Si_3O_8)} \rightarrow$$

$$\underset{\text{potassium ions}}{2K^+} + \underset{\text{kaolinite clay}}{Al_2Si_2O_5(OH)_4} + \underset{\text{silica}}{4(SiO_2)}$$

The potassium ions released from the feldspar may be carried away in solution, utilized by plants, or become incorporated into clay minerals. A small part of the silica is removed in solution, although the greater part remains in the clay-rich weathered residue.

Carbonation

As implied by the term, carbonation involves the chemical addition of carbon dioxide to earth materials. Carbon dioxide in the atmosphere (and in the air trapped within soils) is readily absorbed bo water to form carbonic acid. Although relatively weak, carbonic acid nevertheless has a pervasive cumulative effect in the chemical weathering of a variety of different kinds of rocks. It is involved in the dissolution of common silicate minerals and is particularly effective in dissolving limestones and dolostones. The reaction for limestone is indicated below.

H_2O	+	CO_2	→	H_2CO_3
water		carbon dioxide		carbonic acid
H_2CO_3	+	$CaCO_3$	→	$Ca(HCO_3)_2$
carbonic acid		calcite (in limestone)		soluble calcium bicarbonate

In order for carbonation to occur, water must be readily available. For this reason carbonation is most vigorous in most climates.

In the weathering of silicate minerals, carbonation and hydrolysis work together as the earth's most important processes for achieving decomposition of rocks. The hydrolysis component provides clay minerals and takes silica into solution, while simultaneously carbonation removes metallic elements as ions in solution.

Oxidation

Oxidation, the addition of oxygen to a compound, is one of the main kinds of changes produced in rocks by chemical weathering. Oxygen has a strong affinity for iron, which may be present in such silicate minerals as hornblende, and olivine, as well as sulphides such as pyrite (FeS_2). The oxidation of the iron (essentially what we call rusting) takes place chiefly in the presence of atmospheric moisture and results in the range of red and brown colourations we see in soils, and weathered rocks. In the oxidation process, oxygen gas dissolved in water reacts with iron to from hematite (Fe_2O_3) or limonite ($Fe_2O_3 \cdot H_2O$). The process is illustrated by the following formula:

$$4Fe + 3O_2 + nH_2O \rightarrow 2(Fe_2O_3)\ nH_2O$$

iron oxygen water "limonite" iron hydroxide or "rust")

(*n means a variable amount)

The oxidation of a common nonsilicate such as pyrite (fool's gold") involves combining oxygen with both and sulphur as follows;

$$4FeS_2 + nH_2O + 15O_2 \rightarrow$$

pyrite water oxygen

$$2Fe_2O_3 \cdot nH_2O + 8H_2SO_4$$

"limonite" sulphuric acid

Here, the relatively insoluble iron compounds may remain as a coating on the rock, whereas the sulphuric acid is leached away and becomes available for chemical reactions with other minerals.

Hydration

Hydration is a process whereby water is absorbed by a mineral and incorporated into the weathering product. For example the mineral anhydrite ($CaSO_4$) may take in water to become alabaster gypsum ($CaSO_4 \cdot nH_2O$), or hematite (Fe_2O_3) may be converted to limonite ($Fe_2O_3 \cdot nH_2O$). Hydration is an important process in the development of clay and accounts for the presence of water within many clay minerals. Another aspect of hydration is that the hydrated mineral, because of the water it has taken up, larger than the parent mineral. The increase in volume causes growing hydrated crystals to exert pressure on the walls of the spaces they occupy, and such pressure may contribute to rock disintegration.

Solution

We have seen how the dissolving power of water for certain rocks and minerals is increased when carbon dioxide is present. Even without the addition of car bon dioxide, however, water has the ability to dissolve rocks and minerals. The dissolving away of thick beds of salt and gypsum is an example of simple solution weathering. A quartz sandstone also may weather by solution, although the process is exceedingly slow because of the low solubility of quartz.

The ability af water to dissolve substances is related to the configuration and electrical properties of the water molecule itself. A water molecule consists of a large strongly negative oxygen ion and two hydrogen ions. The hydrogen ions have contributed their electrons to the molecule and thus exist as two small positively charged protons. Both of the protons are located on the same side of the molecule, so that it has an electrically positive side and an electrically negative side. It is termed a dipolar molecule. The electrical charges at either end of the water molecule not only attract their opposites in other water molecules but also attract compounds in rocks and minerals that likewise have separate centers of positive and negative electrical charge. These compounds are taken into solution as weathering proceeds.

Biochemical Weathering

Rooted plants also play a role in the weathering of materials at the surface of the earth. During photosynthesis, plants utilize the energy of the sun's radiatian to combine water and carbon dioxide and produce tissue necessary for growth and maintinance. Part of the hydrogen from moisture taken in by the plant is released at the surface of rootlets where it exchanged for ions of calcium, magnesium and sodium from adjacent particles of clay and other minerals. These exchanged ions are required by the plant for nutrition. The hydrogen ions transferred to the clay particles render the play slightly acidic, and this triggers weathering reactions with neighbouring feldspars and other silicates. Eventualty, some of these minerals also decompose to clay, thus perpetuating the weathering process.

RATES OF WEATHERING

The Role of Climate

How rapidly a particular exposure of rock will weather is dependent on several different interacting factors in the environment. Climate is important because in controls temperature and rainfall. Warmer temperatures, of course, promote chemical reaction. Even an increase of as little as 10°C can double reaction rates. Similarly, high rainfall accelerates rates of weathering by increasing the possibilities for solution and reaction and by helping to wash away surface debris and expose fresh surfaces for chemical attack. Both higher temperatures and an abundance of moisture promote the growth of plants, and luxuriant plant growth assists in weathering by extracting ions from minerals and adding chemically reactive compounds. The importance of temperature and rainfall is at least partly evident in the thickness of loose material or regolith that develops over bed-rock under various climatic conditions. In temperate regions having level topography, the layer of regolith is usually only about a meter deep, whereas in similar topography of the tropics, the loose material may from a blanket tens of meters thick. Arctic areas are only thinly covered by regolith.

Parent Rock

Depending on their composition, texture, and such features as fractures and joints, rocks exposed at the earth's surface may exhibit a wide range of resistance to weathering. Limestones and dolostones, for example, are highly susceptible to solution. In regions having high rainfall, these rocks weather more rapidly than in arid regions. This observation suggests that it is difficult to list rocks in the order of their resistance to weathering because a rock might be more durable than another under one set of conditions and the same rock less durable under different conditions. As a general rule, igneous and metamorphic rocks are more resistant to weathering than sedimentary rocks. A silica-cemented quartz sandstone, however, is among the most enduring of rock types.

Statues and other sculptured edifices that have actual dates carved into the stone permit one to assess rates of weathering. In cities, because of the higher levels of carbon and sulphur dioxides, chemical weathering rates are generally higher than in nonindustrialized areas. The bust of Beethoven was erected in a St. Louis park in 1884. In spite of many attempts to treat the Italian Carrara marble from which the bust was carved, it continued to decompose. The composer's gruff expression has been so altered that St. Louisans refer to the statue as "poor Beethoven."

The texture of the parent rock will also influence its susceptibility to weathering. In general, coarsergrained porous varieties of rocks of the same composition tend to weather more rapidly than dense finegrained varieties. Rocks that have a large number of fractures provide a greater amount of surface area for chemical attack and therefore are also more readily weathered.

Acid Rain

Rates of chemical weathering are, to be sure, affected by the acidity of the solutions that attack rocks and minerals. As described above even unpolluted rainwater contains carbon dioxide and is slightly acidic. In recent years, however, this weak natural carbonic acid has been augmented by strong acids formed where rain has fallen through air polluted with oxides of nitrogen and sulphur. This acidic precipitation includes both nitric and sulphuric acid. It is called acid rain. Acid rain commonly exceeds the natural acidity of rainwater by more than 200 times, and in a few instances by over 1000 times. If not neutralized by alkaline rocks and soils, the acid accumulates in lakes and streams where it is often lethal to fish and other forms of aquatic life. Acid rain also damages the foliage of plants and reduces the quality of soil by leaching away nutrients. The acid solutions attack limestone, marble, and concrete building materials and damage valuable works of art. Presently, millions of dollars are being expended in acid rain research in the hope that a solution to the problem can be found. Aside from local chemical treatment of particular areas, however, a long-term solution must involve reduction in atmospheric pollutants that form acid rain.

SPHEROIDAL WEATHERING

Spheroidal weathering is a term used to describe the spalling away of concentric surphicial shells of the rounded surface of a boulder or rock mass. Such "onion-skin" weathering its believed to result primarily from the mechanical effects of chemical weathering. When feldspars decompose the clay product has a greater volume than the parent feldspar. The increase in volume disrupts the interlocking texture of mineral grains in the rock and causes breakage and separation of the layer of partially weathered rock near the surface. Corners of roughly rectangular blocks broken loose along joints are decomposed along three surfaces simultaneously and are progressively rounded by spalling until the rock mass assumes the spheroidal form.

Spheroidal weathering operates most effectively on relatively small rock masses such as those of boulder size. The kind of spheroidal weathering involving a larger-scale breaking off of concentric plates from bare rock surfaces is referred to as exfoliation. Rocks that have experienced exfoliation may resemble those having sheet structure, but while sheet structure is caused by release of pressure when overlying rock masses are removed by erosion, all forms of spheroidal weathering result from physical and chemical weathering.

THE DECAY OF GRANITE

The mineral alterations involved in chemical weathering are nicely illustrated by the decay of a granitic parent rock. Granite is a common igneous rock composed of about two parts orthoclase, one part quartz, one part plagioclase, and small amounts of ferromagnesian minerals. Where masses of granite have been exposed for a long period of time, one may find places where the weathered products of the granite have not been washed away but have accumulated as a clayey granular residue. On close examination, this material is found to consist quartz grains, partly decayed and clay-coated feldspars, rust-coloured particles of partially decayed ferromagnesian minerals, and clay. The quartz grains in the weathered material are relatively unaltered because of the great resistance quartz has to chemical attack. As other minerals in the granite decompose around them, quartz grains are freed from the rock matrix, later to be transported and deposited as sand, and perhaps ultimately to become components of sandstone. Granite provides a good example of differential weathering, by which its various components weather at different rates.

We have noted that feldspars are aluminosilicates of potassium, sodium and calcium In the weathering process, potassium, sodium, and calcium are largely dissolved and carried away in solution. Subsequently they may be combined with other elements and incorporated into sedimentary rocks. At least some of the potassium is retained in newly forming clay minerals. The remaining aluminum and silicon in the feldspars become the chief ingredients of clay and this accounts for the coating of clay frequently found around feldspar grains and the clay residue found adjacent to bodies of weathering granite. Later, this same clay may find its way into the making of sedimentary rocks like shale and claystone.

During the decomposition of the ferromagnesian minerals present in the granite source rock, potassium, sodium, and calcium are dissolved in the same way as in the fieldspars. Again, aluminum and silicon go into the construction of clay minerals. The ions that remain are iron and magnesium. The iron combines readily with oxygen to form iron oxide minerals such as hsmatite (Fe_2O_3) and hydrous iron oxide minerals like goethite, FeO(OH). These iron minerals colour sediment and rocks in tints of yellow, orange, and brown. Finally, the magnesium that is derived from parent ferromagnesian minerals may find its way into limey sediments or become a component of certain clay minerals.

SOIL

In terms of importance to human populations, the most significant aspect of weathering is in the development of soil. Soil can be defined as weathered material that will support the growth of rooted plants. A heap of quartz sand is not soil according to this definition, for it will not support vegetation. Nor is the so-called "lunar soil" of the moon a true soil. It is merely disintegrated mineral matter and should more properly be termed regolith. Regolith refers to the debris of rock and mineral matter that rests on solid bedrock. It includes soil, alluvium, and weathered fragments of the bedrock.

Soil is not a mere accumulation of particles derived from mechanical and chemical weathering. It is a vital natural material that would not develop were it not for the activities of a myriad of bacteria, fungi, worms, and insects. Actually, soil can be considered an intricate mixture of mineral solids, water, gases, dissolved substances, remains of dead organisms, and multitudes of thriving organisms.

Why is a certain amount of organic material necessary in order for weathered material to support the growth of rooted plants? One reason is that the organic material supplies nutrients required for vigorous plant growth. Plants need nitrogen in order to synthesize protein, but they are unable to take free nitrogen (N_2) from the air. Certain soil bacteria, however, are able to use this free nitrogen and convert it into usable fixed nitrogen compounds (nitrates). Also, the decaying vegetation and animal waste contain ammonia (NH_3) the nitrogen of which can be utilized by most plants.

The material that colours the upper parts of many soils gray or black is called humus. Humus is organic material that is so thoroughly decayed that one cannot discern the nature of the parent organisms. This essential organic mixture increases the prorosity and water-holding capacity of the soil, provides a buffer against rapid changes in acidity, and assists in the retention of chemicals needed as plant nutrients. Carbon dioxide liberated by humus will, as we have noted, combine with water to form a weak acid capable of dissolving calcium carbonate. Once the calcium is separated from the carbonate part of a compound, chemicals in humus known as chelating agents, bond to the calcium and prevent it from again forming $CaCO_3$. Most plants are unable to extract the calcium they need directly from $CaCO_3$, but they have evolved mechanisms for releasing the chelated calcium that exists in humus.

Factors Governing Soil Development

Five factors are involved in soil formation. These are climate, parent material, topography, time, and biologic activity. Of these factors, climate is of greatest importance. Indeed, soil types and climate are so closely related that maps showing global distribution of climates resemble maps showing the occurence of different soil types. Temperature and rainfall, of course, influence not only rates of weathering but the nature of vegetation. Given enough time, soil-informing processes operating within a given climate will prevail over differences in parent material and provide a basically climate-controlled soil type. The reserve is also true. Identical rocks in temperate high-rainfall regions and in semitropical arid regions will produce quite different soils.

The influence of parent material on the formation of soils is of secondary importance but can be observed in soils that have recently developed from unaltered bedrock. Such soils stills retain textural and mineral components that are derived from the parent rock but which may become obscured after the soil has evolved to a further stage. In some cases, various influences of parent material will persist in the developing soil. Soils develaped above quartz sandstones may tend to be more acidic than soils formed on limestones, and these differences will be reflected by the natural vegetation living on each soil type. Soils forming on rocks that are deficient in certain elements such as magnesium or boron will ordinarily also lack these constituents. Not all soils are developed from underlying consolidated bedrock. Many fine agricultural soils are formed on loosely consolidated sediment laid down by streams, winds, glaciers, or the waters of lakes.

Topography also influences the ultimate nature of soils. On steep slopes, for example, erosion is usually more vigorous and the soil layer generally thinner. Also, water falling on slopes is partially lost to runoff, and therefore there is less water available for percolation into deeper layers of soil. Even the direction of slope may have a bearing on soil development. In the northern hemisphere, slopes facing toward the south tend to be warmer and dryer and therefore support quite different kinds of plants than do the northern slopes.

The factor of time in soil formation is often not fully appreciated by those who clear away soil for construction or mining and expect nature to quickly restore this important resource. The development of soil is an exceedingly slow process. Depending on climatic conditions, several hundred to several thousands of years are required to produce a fertile soil. Agricultural soils formed during this lengthy natural process may be lost to erosion in only a few years, particularly in areas where soil conservation measures are not followed. The problem is not trivial, for an estimated 3 billion metric tons of good agricultural soil is lost each year from agricultural fields in the United States alone.

Soil Horizons

Soils are not uniform in texture and composition from bedrock to the surface but rather consist of a number of fairly distinct ayers or soil horizons, which differ in composition and physical characteristics. The stack of soil horizons is called a soil profile. A simple profile, such as might develop in a humid region on granitic rock, has three distinct divisions. Soil scientists have named these layers the A, B, and C horizons. The A horizon, also known as "top soil," which lies immediately beneath the surface, is characterized by a high content of organic matter. It is also a zone in which soluble compounds are dissolved and, along with fine clay particles, carried downward to be deposited in the underlying B horizon. Becauses of these losses, the A horizon is referred to as leached or eluviated, whereas the B layer is sometimes called the washed in or illuviated zone. Iron and clay minerals tend to accumulate in the B layer. In humid regions one can often recognize the B horizon by its more brownish colour and clayey texture. The C horizon of the soil profile consists of partially altered parent material. In it one finds evidence of chemical weathering, but soil development has not progressed to the level at which the original characteristics of the bedrock are unrecognizable. Beneath the C horizon lies unaltered parent material.

Soils developed on limestones in humid regions often do not exhibit a clear division between the B and C horizons. In such soils, the A horizon consists of dark humus-rich material that changes downward into relatively light coloured clay and then stained and partly dissolved limestone. The clays in these soils are residues of clay impurities left when the limestone was dissolved.

How does the sequence of soil horizons develop in a humid region underlain by granitic rock? Imagine that glacination has recently stripped away all loose sediment over a granite outcrop. Initially, disintegration and decomposition of the granite will produce a sandy layer composed mostly of grains of quartz the feldspar along with clay derived from the weathering of silicate minerals. As the layer of granular material thickens, plants establish themselves, and organic material from a variety of sources becomes incorporated into the surface layer. Meanwhile, with each rainfall, soluble ions and clay particles are flushed downward from the surface layer into a lower level. This lower level soon takes on the clayey character of the B horizon. While these events are taking place, solutions reaching still deeper continue the chemical attack on the parent material, thus perpetuating the C horizon.

Major Kinds of Soils

Soil scientists employ a rather complex nomenclature for the great variety of soil types present at one place or another around the globe. Fortunately, most of the names in these lengthy classifications can be avoided if one desires only a general understanding of a few major kinds of soils. For example, the soils that predominate in the eastern United States and southeastern Canada are called predalfers. The name was fabricated to emphasize the fact that in such soils aluminium (al) and iron (fer) have been leached from the A horizon and deposited in the underlying B horizon. Pedalfers are clayey soils that develop in regions having an abundance of rainfall. As this water percolates through the humus, it becomes acid, and this accounts for its ability to leach aluminium and iron from the topsoil. It is also effective in dissolving carbanates, which are then carried away by ground water. The acidity of some pedalfers may diminish their fertility, and so farmers often spread finely ground limestone (agricultural lime) on their fields to combat the actidity.

As a consequence of the fact that the top soil or A horizon in pedalfers is leached, it is usually a lighter colour than the underlying subsoil where iron has accumulated. Pedalfers include several lesser categories of soils, which differ in their profiles. Most important among these are podzols, which typically have an ashy gray A horizon.

In the more arid regions of the world, less humus develops in soils, and there is less opportunity for solution and leaching. In the absence of water, chemical, weathering is slower, and less clay is produced. Usually, there is insufficient ground water to flush soluble materials such as calcite out of the B horizon, so it accumulates there as soil moisture is lost by evaporation. Because of the persistence of calcite in these soils, they are called pedocals (pedon, soil, and call for calcium carbonate).

Frequently in dry areas, there is not only insufficient water to cause downward leaching, but there is an upward movement of water because of the high rate of evaporation at the surface. Mineral matter dissolved in the diminishing soil water is precipitated at the surface as a hardpan or caliche layer.

In the hot and humid regions of the tropics, the characteristic soils are laterites. The term "laterite" is derived from the Latin word latere or brick, and originally referred to the use of this material to make bricks in India and Cambodia. Laterites have a telatively thin organic layer covering a reddish leached layer, which is often underlain by a still darker red layer. In laterites, oxidation and hydrolysis have been so intense that feldspars and ferromagnesian minerals are completely decayed. Not only is calcium carbonate removed, but also silica. Only the most insoluble compounds, mainly aluminium and iron oxides, accumulate in these soils.

Although laterites may support a lush growth of natural tropical vegetation, they are not good agricultural soils. When forested areas with lateritic soils are cleared and plowed, the thin organic cover tends to be rapidly oxidized in the prevailing warm climates. A thick accumulation of organic matter such as that found in rich black soils of more temperate regions cannot develop. After a few years of tilling and planting, the organic component of the soil is so depleted that fields must be abandoned.

In some laterites, the concentration of either iron or aluminium may reach levels that permit the deposits to be profitably mined. The iron-rich laterites originate from the weathering of parent materials that also contained iron, although not concentrated into ore bodies. Extensive lateritic ores of iron occur in Cuba, Columbia, Venezuela, and the Philippines. If there is little iron in the parent material and an abundance of aluminium, then laterites rich in hydrated aluminium oxides map form. Such materials are called bauxites. Ancient bauxite deposits are mined in Guyana, Ghana, northern Queensland, and Arkansas. Concentration of bauxite takes place in tropical areas of low relief where temperatures exceed 25°C most of the time and where there is an abundance of water for leaching and chemical reactions. At the present time, bauxite is the only are from which it is economically practical to extract aluminium.

Soil Erosion

It takes nature thousands of years to produce a fertile soil. That same soil can be lost to erosion in only a few decades. Clearly, as we attempt to feed our expanding world population, care must be taken to prevent unnecessary crosional losses of this vital resource.

Soil erosion results primarily from uncontrolled runoff of surface water. In areas that have never been tilled, soil is protected by a cover of plants, and there is an approximate balance between the slow loss of soil by erosion and the development of new soil from underlying parent material. Whenever the protective shield of vegetation is broken, however, erosion will follow. There, are, of course, natural events that can destroy the vegetative cover. These include droughts, plagues of insects, epidemics of plant disease, and fires caused by lightning.

Agriculture, however, has been a far more potent factor in causing erosional loss of soils. Some of the erosion resulting from farming is unavoidable. We must have food, and so we must break into the natural vegetative cover. There are, however, farming methods that reduce the loss of valuable topsoil to erosion, and these methods are practiced widely. For example, farmers now plow and plant their crops in rows that follow contours so that rainwater is retained rather than allowed to run off. Efforts are made to reduce the amount of time between preparation of the soil for planting and planting itself so that the bare soil will not remain exposed to erosion for long periods. Strip-cropping is also practiced. This technique involves planting alternate bands of erosion-resistant crops such as clover and alfalfa with open-spaced crops like corn. Soil losses in the erosion-susceptible, corn strip are trapped in the adjacent strip of denser crops. Where fields are temporarily not utilized, soil-holding "cover" crops are planted, not only to retard erosion but to improve the soil by adding nitrogen. Because ordinary grass in an effective retardant for erosion, overgrazing by cattle and sheep must be prevented.

One of the most troublesome erosional problems faced by farmers is the control of gullies. Gullies start in either natural depressions or plow furrows and generally extend themselves toward higher ground by a process called headward erosion. The gully head receives rainwash from a wide area. This water converges so as to flow rapidly into the steep depression at the upper end of the gully, eroding vigorously as it enters the channel. It is this rapid erosion that causes the gully to cut ever farther in the headward or upslope direction. To halt this destructive process, steps can be taken that prevent water from entering the gully. Normally this is done by excavating a shallow diversion ditch around the gully head to trap its potential water supply and "starve" the gully. Because gullies grow by headward erosion, disposing of old refrigerators and automobiles in these ditches has no effect at all in retarding their growth.

3

Roots and Soil

All nutrients and water which enter plants traverse the interface between roots and the soil; yet this boundary has usually been little considered in discussions of the physiology of plant growth or of soil science. It might be described as a 'no-mans' land between them. Except when the presence of mycorrhizas compels attention, the abundant microbial flora which plant exudates support has usually been assumed neither to accelerate nor retard the flow of mineral nutrients to plant roots; similarly the impression, which can be encouraged by a superficial inspection, that roots and root hairs make uniform and continuous cantact with the bulk soil, has been widely accepted without question.

The adequacy of this simple model is examined in this chapter. Uncertainties which beset present evaluations of factors which control the transfer of water and nutrients from the bulk soil to roots especially under field conditions are emphasised. The possible role of root hairs is discussed.

DEFINITION OF THE INTERFACE AND ITS NATURE

Narrowly defined, the term 'soil/root interface' could imply merely the boundary between the outermost root tissues and the particles of soil which are closest to them. But a much broader definition is necessary to consider adequately the extra-cellular resistances which may influence the transfer of water and nutrients to roots from the 'bulk soil'; the term 'bulk' soil is used here as to denote the zone of soil which lies outside the rhizosphere zone and is not directly influenced by growing roots except by the withdrawal of water and nutrients. The principal biological and physical characteristics of the interface can be summarized as follows.

Biological Aspects

The walls of the outermost roots cells form the inner boundary of the interface. In relatively young roots, such as those of annual plants, these tissues are the epidermis and root hairs. But discussion of the role of root hairs in absorbing water and nutrients from soil was deferred until the concluding section of this chapter. The other principal biological components of the interface have already been described:

—mucigel which can create intimate contact between the root or root hairs and adjacent soil particles.

—soluble exudates.

—the microbiological flora of the rhizosphere and rhizoplane, the substrates of which are mucigel; soluble exudates and, in some instances, the cellular contents of living plant tissues.

Physical Aspects

From the viewpoint of the transfer of water and nutrients to plants, one of the most important physical characteristics of the interface is the degree of physical contact between root and soil. If this contact were significantly less than that between components of the bulk soil an additional resistance to the movement of water and nutrients could be created. Do such situations arise and if so can they be of major significance in determining the performance of root systems?

The simplest situation, and that which has usually been alone considered in detail is when young roots are penetrating finely divided soil which is adequately supplied with water, as for example, in laboratory pot culture experiments. Superficial inspection of roots grown in this way can create the impression that they are in continuous contact with the soil, but microscopic examination often reveals appreciable discontinuities even with young roots, when the cortex is intact (see, for example, Drew and Nye, 1969).

Direct evidence is meagre on the degree of contact between roots and soils in more natural circumstances. The difficulty of preparing sections of soil which contain roots for microscopic examination, without distorting either the cavities or the roots they contain, has discouraged the extensive examination of this question. However, Sheikh and Rutter (1969) found that the diameters of the roots of a sub-shrub Erica tetralix, and a grass Molinea caerulea, growing in a natural habitat in southern England were commonly less than half of that of the cavities surrounding them. Additional information comes from comparing pore space in soil with the dimensions of root systems. The observations of Welbank *et al.*, (1974) suggest that at mid-season the roots of winter wheat occupy less than one per cent of the volume of soil in the upper 15 cm in which the density of roots is greatest; this figure is derived on the assumption that their dry matter content is *c.* ten per cent. Table 2.3 however shows that pores exceeding 300 μm in diameter, which appears to be an approximate upper limit for the mean diameter of well established cereal root systems (Hackett, 1969), often exceeds two to five per cent of the total volume of many, though not all, types of soil; appreciably higher figures are sometimes reported. Accordingly, bearing in mind that roots tend to proliferate in soil where they can penetrate most freely, it appears likely that appreciable parts of root systems often develop in pores of diameter larger than themselves. The same conclusion is encouraged by the distribution of roots in soils of appreciable strength and by observations of roots against panels in root laboratories, quoted by

Tinker (1976). Their preferential growth in voids or planes of weakness between structural units can be readily observed; in such circumstances roots may not be in contact with the soil over their entire surface. Accordingly, despite the lack of extensive quantitative information, the assumption that growing roots normally make continuous contact with the soil seems questionable. Beyond this, when roots have entered pores, contact between them and the soil may sometimes be impaired either by the enlargement of the cavities or by the decreasd in diameter of the roots themselves.

Soil characteristics which can modify the diameter of cavities occupied by roots. Soils which contain an appreciable quantity of clay shrink on drying. As the growth of roots is dependent on an adequate water supply, young roots are likely to grow in soil which is at or near its maximum expansion. Loss of water from the soil due to dry weather and to its removal by roots may thus cause the soil to contact as growth proceeds; sometimes shrinkage and the consequent cracking of the soil are indeed sufficient to ruptureroots. A mechanism which could increase the diameter of voids occupied by roots is thus apparent.

Characteristics of growing roots which can modify their diameter. In dicotyledons the diameter of roots may increase with time due to cambial activity; other factors being constant this should reduce any gap between the root and the solid phase. This progressive thickening does not occur in monocotyledons as they lack a vascular cambium, and the diameter of their roots can decrease as they age because of the degeneration of the cortex due to invasion by micro-organisms or desiccation.

Investigations by Huck *et al.*, (1970) and Faiz (1973); see also Weatherley (1975) have lately demonstrated a further important response of roots. Huck, *et al.*, (1970) studied diurnal changes in the diameter of individual roots of cotton plants (Gossypium hirsutum) through a glass panel in the Auburn rhizotron). Figure 2.1 shows measurements made over a period of four days during which the water potential of the soil decreased and the net solar sadiation increased daily so that increasing water stress was experienced. On the first day the diameter of the root remained relatively constant but on the second some diurnal change was apparent. This increased as the experiment proceeded and on the last day, some three hours after the peak radiation load, the root shrank temporarily to about 60 per cent of its average diameter. Pictures of one root at different times of the final day are reproduced.

Faiz (1973) found a comparable effect with sunflower (Helianthus annuus) roots grown in pot culture. Exposure to an atmosphere of 40 per cent relative humidity for nineteen hours caused the leaf water potential to drop from —6 to below —12 bar; this change is within the range found under field conditions in dry wheater. At the end of the drying cycle the diameter of the roots was about 75 per cent of that at the beginning. These results are readily compatible with the more familiar observation that the diameter of the above-ground tissues can be reduced by water stress (Kramer, 1969; Klepper *et al.*, 1971). The effect on the diameter of roots appears to be of major interest in relation to the absorption of water, since it suggests that contact between roots and the soil can be

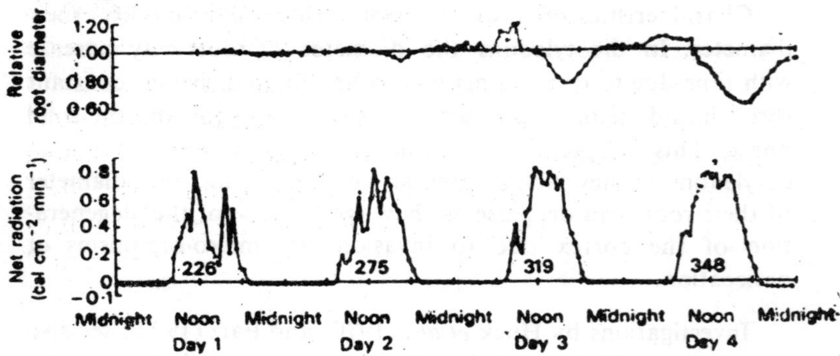

Fig. 2.1. Variations in root diameter and net radiation during the course of four consecutive days. Numbers under peaks are total net radiation during daylight

impaired under dry conditions when the ready transfer of water across the soil/root interface is particularly important for maintaining leaf water potential.

The complexity and variability of the situation at the soil/ interface is thus indicated by several types of evidence; physical contact cannot be assumed to be necessary continuous or complete. But, by itself, this does not establish that the interface exercises a major control over the movement of water and nutrients from the soil into plant roots; this would be so only if the resistances at the interface were large relative to those in the bulk soil or within the plant itself. In the sections which follow this question is considered in relation to the movement of water and of nutrients.

THE TRANSFER OF WATER ACROSS THE INTERFACE

The resistance of the soil to the movement of water to roots is often regarded as consisting of two consecutive components—the pararhizal and the rhizosphere resistances. The former, which lies outside the scope of this book, is the resistance to the movement of water through the bulk soil to the rooting zone, the rhizosphere resistance being that in the soil immediately adjacent to each single root (see Newman, 1969a). This terminology implies that the latter resistance resides entirely in the soil itself and takes no account of possible variations in physical contact between root and soil. Thus, the term soil/root interface resistance', or for brevity interface resistance seems more appropriate and is here used.

Interest in the interface resistance has hitherto centred largely on its implications with regard to the effect of root density on water uptake. If the interface resistance were sufficiently high the roots of a rapidly transpiring plant might reduce the water potential of the neighbouring soil to near the wilting point even though the water potential a short distance away was close to field capacity; roots would thus have access only to water in their immediate vicinity and the quantity of water they could extract would be closely related to the total density of roots, *i.e.*, their length per unit volume of soil. If, on the other hand, the interface resistance were relatively low, water potential would change only slowly with the distance from roots and the zone from which they could draw water would be considerably larger; root density would be less important in determining water uptake.

Theoretical Evaluations

Current experimental methods do not enable differences in water content or in water potential to be measured over small distances in soil which are comparable to the diameter of roots. Estimates of gradients of water potential in the near vicinity of roots thus rest on theoretical evaluations; those of Gardner (1960) and Cowan (1965) have been followed by numerous subsequent studies. In brief the basis of these calculations is:

1. The hydraulic conductivity of the bulk soil is measured over a range of soil water contents.
2. Uptake per unit length of root is calculated from the dimensions of the root system and the measured rate of transpiration, it being normally assumed that water potential is initially uniform in the bulk soil and that water enters all parts of the root system equally. No account is thus taken of the gradients in water potential commonly observed in the field.

Figure 2.2 shows the relationship calculated by Cowan (1965) for the decrease in water potential in the vicinity of a root 1 mm in radius into which water enters with a velocity of 2.5 mm d^{-1} (equivalent to the uptake of c. 0.16 cm^3 d^{-1} per cm root length) which was regarded as being a high value but within the range calculated from laboratory experiments. Cowan's results suggest that the soil water potential increases only gradually with distance from the root unless the matric potential of the bulk soil is below —4 to —5 bar. Gardner

(1960, 1965) had reached broadly similar conclusions. These analyses justified the abandonment of the earlier assumption that the interface resistance would cause the movement of water towards roots to be negligible whenever the neighbouring soil was appreciably below field capacity. It is important to recognize, however, that if the uptake of water per unit length of root were greater than that assumed in the calculation the gradient in water potential would be correspondingly steeper.

Following the work of Gardner and of Cowan, Newman (1969a) estimated the soil water potential and root densities at which the resistance to water flow across the soil/root interface might be expected to have an appreciable effect on absorption; he assumed that this would happen when soil resistance reached 25 per cent of the plant resistance. Transpiration rates,

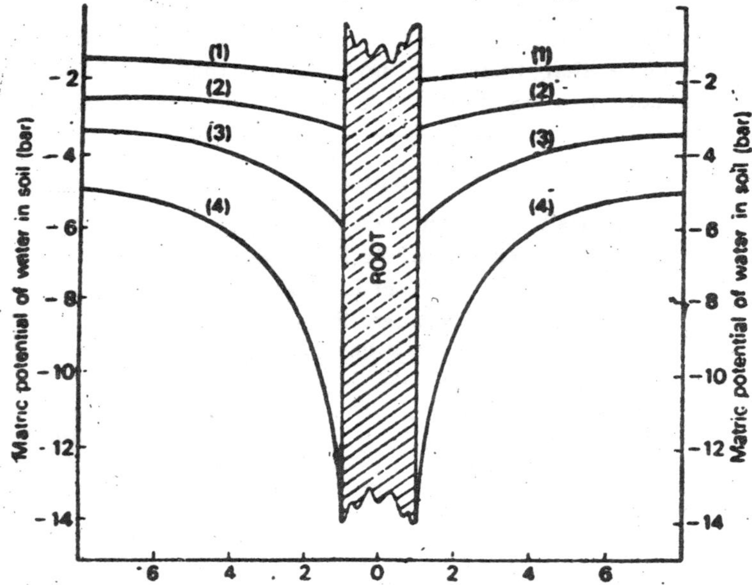

Fig. 2.2: Calculated gradients of water potential in the neighbourhood of a root 1 mm in radius which is absorbing water at the velocity of 2.5 mm day^{-1}. The curves refers to the following concentrations of water 8 mm from the root (1) 0.20; (2) 0-185; (3) 0-17: (4) 0-155 cm^{3} em^{-3} soil

judged as representative from the literature, and a constant value for plant resistance, namely that assumed by Cowan, were used. The water potential was regarded as uniform throughout the rooting zone. Newman concluded (Table 2.1) that unless the total length of root present in soil per unit surface area was less than 200 cm cm^{-2} the resistance to water flow across the interface would impose a minor restraint on water uptake except when the matric potential of the soil was below—16 bar. Since root length for many species is considerably higher than 200 cm cm^{-2} Newman concluded that root systems were usually much more extensive than necessary for water uptake. This was compatible with evidence he had previously obtained from experiments with plants grown in pot culture (Andrews and Newman, 1968); considerable root pruning did not increase drought susceptibility. Newman (1969b) subsequently re-interpreted the results of a number of other investigations, mainly with plants grown in shallow containers, and judged that there was no evidence that the 'interface resistance' was significant except when the matric potential of the soil was relatively low, *e.g.*, below —7 bar.

TABLE 2.1

Estimates by Newman of the Soil Water Potential at which Resistance of the Rhizosphere Soil will Reach 0.25 of Plant Resistance if Varying Lengths of Root are Present in the Soil Per cm^2 Soil Surface

Length of roots in soil (cm per cm^2 soil surface)	*Soil water potential at which resistance in rhizosphere soil equals 25 per cent of plant resistance (bar)*
1	—0.7
10	—2
50	—6
100	—10
200	—16
1000	> —40

The interruption of contact between roots and the soil has seldom been considered in theoretical assessments. Cowan and Milthorpe (1968), however, deduced that the gap between the root and the soil seemed likely to have little effect on the movement of water because of its transfer in the vapour phase.

Studies of Plants Subject to Water Stress

What is the relevance of these deductions to the response of crop plants subject to water stress under natural conditions? Both Cowan and Newman considered situations in which water entered all parts of the root system at a relatively uniform rate; there is no reason to doubt that this can happen if the water supply and other factors are similarly constant throughout the rooting zone. However, practical interest in water uptake centres on situations when plants growing in the field are subject to water stress and this commonly occurs because of low water potential in the upper soil layers, plants are then dependent on absorption by the deeper root members which in the majority of crops often represent a small fraction of the total root system. These conditions cannot be simulated when root systems are confined in small containers; Taylor and his colleagues (Taylor and Klepper, 1975; Browning *et al.*, 1975) have conducted some of the only detailed studies which were not subject to this limitation. Cotton *Gossypium hirsutum*) was grown in large rhizotron chambers filled with uniform fine sandy soil in which a natural profile of increasing water content from the soil surface downwards could be simulated. Their results show that water uptake per unit length of roots increased considerably, sometimes more than ten-fold; with increasing water stress. They developed a more direct experimental approach to studying resistances to the transfer of water from soil into roots. Water uptake per unit length of root was calculated from changes in soil water and measurements of root length. The water potential in the root xylem was assumed, as a first approximation, to be similar to that measured in leaves. Thus, they could calculate the hydraulic conductivity of the 'soil/root xylem pathway', *i.e.*, across the soil/root interface and through the cortex to the conducting tissues of the root.

They found that the hydraulic conductivity of the soil/root xylem pathway was much less than that in the surrounding soil, except when the water content of the soil was extremely low (Fig. 2.3). Whereas the conductivity of the bulk soil rose steadily with increasing water content that of the soil/root rxlem pathway changed little when the water content of the

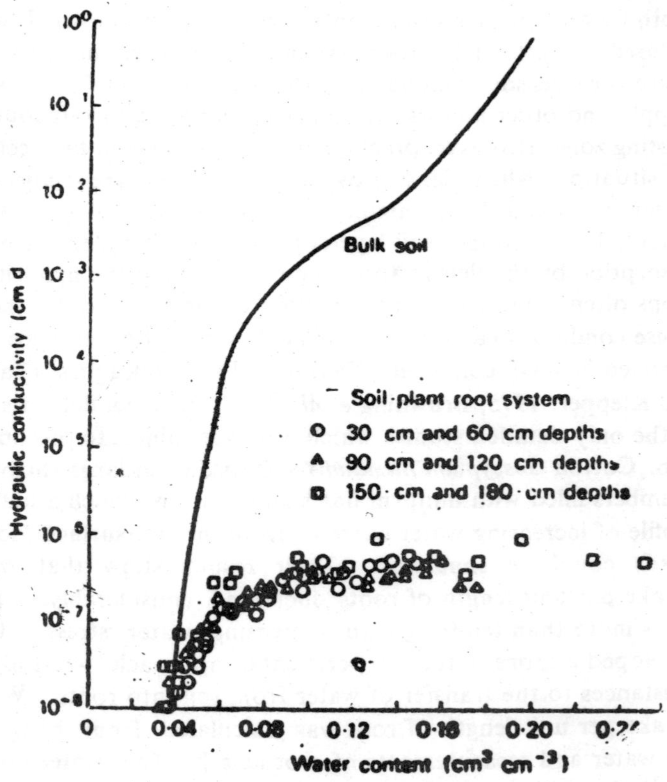

Fie. 2.3: A comparison of the hydraulic conductivities (k) of bulk soil with that of the soilplant root system at various soil depths and soil water contents—results for experiments with cotton (Gossypium hirsutum)

soil was above circa eight per cent dry weight. Since the major resistance to the movement of water was associated with the root the uptake of water per unit volume of soil was proportional to rooting density, other factors being equal. This conclusion is in conflict with the theoretical deductions previously discussed.

Unfortunately little comparable information is as yet available for other species or soils. It is necessary to rely on inference. Information on root distribution shows that a major part of the root systems of crop plants usually develops in the surface soil. Thus, Welbank *et al.*, (1974) found that whereas in June the total length of winter wheat roots per unit area of ground surface of a light soil was c. 200 cm cm^{-2}, only about 30 cm cm^{-2} was below 25 cm, a depth to which water could be relatively rapidly depleted if dry weather ensued so that they alone would be absorbing water. Calculations such as those in Table 2.1, which assume uniform absorption throughot the root system, may thus have little relevance to estimating the size of root systems necessary to sustain plants in a natural environment.

The methods used by Taylor and Klepper (1975) did not enable them to identify the resistances to water movement in its radial transfer from the bulk soil to the root xylem. Evidence reviewed leaves little doubt that the tissues within the root, particularly the endodermis, may make an important contribution, but the observations of Huck *et al.* (1970) and Faiz (1973) and Weatherley (1975) referred to that the diameter of roots can decrease appreciably under water stress suggest that the 'gap' at the root/soil interface may contribute to the overall resistance under conditions of water stress. At present the only quantitative information comes from the work of Faiz who studied water potentials in sunflower plants and the soil surrounding their roots under continuous illumination at a low relative atmospheric humidity. These conditions normally caused the water potential both in leaves and the soil to decrease steadily but when the pots in which the plants were grown were vibrated for one minute on five occasions during an eight hour period the decrease in leaf water potential was checked and the soil water potential decreased more (Fig. 2.4). A comparable result was obtained when plants were grown in plastic containers which were gently squeezed at intervals. It seems reasonable to infer that the effect of both these treatments was because the closure of the root/soil gap facilitated the transfer of water. Faiz (1973) estimated that the reduction in the diameter of roots caused by water stress was accompanied by an increased resistance to water movement at the soil/root interface which was equivalent to about twice the normal resistance throughout the plants. Since the shrinkage of roots occurred only when plants were already subject to water stress this additional resistance was a consequence of water stress and not the primary cause.

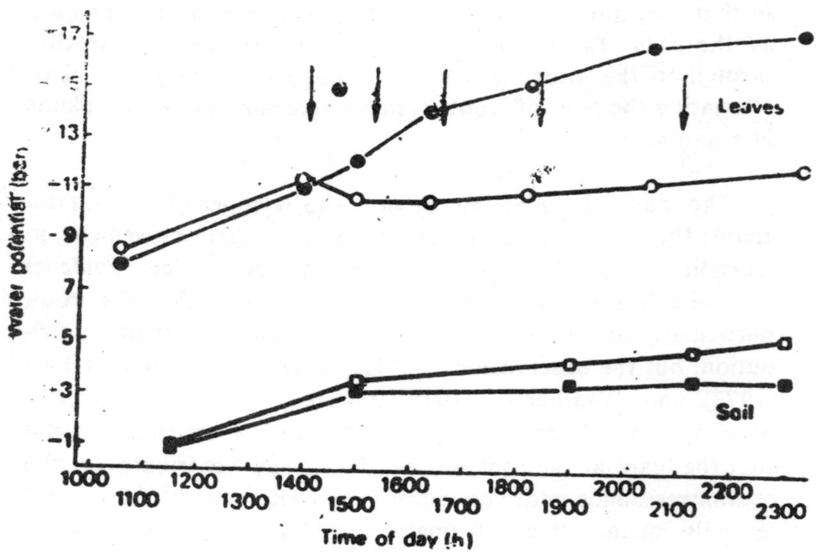

Fig. 2.4: Changes in the water potential in leaves of sunflower (Helianthus annuus) and in soil during a drying cycle when the plant pots were either vibrated briefly on five occasions indicated (open symbols) by arrows, or left undisturbed (closed symbols)

Thus, contrary to the view suggested by some earlier studies, it now appears that inadequate contact between soil and root can sometimes create an important added resistance to the transfer of water across the soil/root interface. However, this effect has been demonstrated only in plants experiencing water stress in uniform mixed, light rooting media. It is tempting to speculate that the effect of root/soil gap may be at least as great in heavier soils with a well developed structure as an appreciable part of the root system may then have developed in the larger voids of channels. This could be an additional cause of impaired contact, but at present there is no information. There can be little disagreement with the conclusion of Tinker (1976) that as yet there it no adequate basis for interpreting some of the relationships at the root/soil interface.

TRANSFER OF NUTRIENTS ACROSS THE INTERFACE

Since the absorption of nutrients is likely to be very restricted if the water potential in the soil is low it is, in practice, relevant to consider possible resistances to ion transport across the soil/root interface, mainly when water is not limiting.

The contrasting rates at which different untrient ions diffuse through the bulk soil to roots was illustrated in Those which move freely in the soil by mass flow in water should experience no greater restraint than water itself in the passage across the soil/root interface. The absence of any significant resistance in the interface to the arrival of sulphate when water uptake is not restricted is evident from the fact that the slower uptake of sulphate than water can cause it to accumulate at the root surface (Wray and Tinker, 1969); this can happen also with calcium (Barber and Ozanne, 1970).

Transfer across the soil/root interface deserves much greater consideration with ions which diffuse slowly in the soil, phosphate and potassium being in practice the most important ones. Their movement through the bulk soil towards roots has been studied in considerable detail and there is convincing evidence that the rate at which they diffuse towards roots can be the rate limiting step in nutrient uptake. Numerous investigations piont to the conclusion that when plants are grown with an adequate supply of light, water and nutrients, and at a favourable temperature in finely sieved soil, the maximum rate of uptake of these ions approximates to the rate at which they could diffuse through the bulk soil to a 'zero sink'; in other words, a rate of removal which causes their concentration to be negligible in soil adjacent to the sites of absorption (Bouldin, 1961; Barber 1962; Passioura, 1963; Nye, 1966; Olsen and Kemper, 1968; Drew *et. al.*, 1969; Nye and Tinker, 1969; Drew and Nye, 1969).

The importance, from this viewpoint, of information of the diffusion of phosphate and potassium has encouraged more detailed studies directed both to interpreting the role of root hairs in absorption (the subject of the subsequent section) and to developing predictive models to describe the depletion of these ions in the soil and uptake by plants. The uncertainties which now be set these endeavours should be recognized; two principal assumptions necessarily underly them at present:

— that the 'root absorbing power' per unit surface area can be regarded as a reliable physiological constant during the course of an experiment;

— that ions diffuse through the soil to the sites of absorption at the rate which can be measured in the bulk soil. This implies that the abundant exudates and organisms in the rhizosphere do not affect diffusion—a question on which direct evidence is lacking.

None the less, some theoretical calculations of the depletion of nutrients in the vicinity of roots, made on these assumptions, agree well with observation. This happened in Bhat and Nye's (1974) studies of profiles of phosphate concentration in the close vicinity of onion (Allium cepa) roots which lack root hairs. However, there were considerable discrepancies in similar studies with rape (Brassica napus) (Bhat and Nye, 1973) irrespective of whether root hairs were assumed to be active or inactive in uptake, considerably more phosphate was withdrawn from the zone external to the root hairs than theory suggested Fig. 2.5). As the authors state, uncertainty in the derivation of the diffusion coefficient for phosphate in the bulk soil could explain this discrepancy, though why it did not also occur with onion is not clear. More recent work (Brewster *et al.*, 1976), to which further reference is made has led to the alternative suggestion that the role of root hairs may be more complex than was previously suggested. Unfortunately, however all the information now available relates to plants growing in uniformly mixed, finely divided soil. The effects of the soil/root interface on nutrient uptake under more natural conditions awaits examination.

A crucial requirement for resolving this matter is information on whether relatively immobile ions diffuse close to the root at the same rate as in the bulk soil. This has been tacitly assumed in the investigations discussed above, but it has as yet not been rigorously tested in natural depends several reasons encourage doubt as to its validity. It is well known that pH at the root surface may differ from that in the bulk soil, the situation being highly variable depending on the ionic balance; Riley and Barber (1971) have suggested that these pH changes may influence the uptake of phosphate. In addition, it seems possible that the accumulation of ions near the root surface (*e.g.*, calcium or sulphate) may effect the mobility of other ions (Hoffman and Barber, 1971).

It is relevant also to enquire whether exudates or microorganisms influence the mobility of ions in the rhizosphere soil. Attention is directed to this possibility because the slow diffusion, characteristic of some ions in the soil, is due to their interaction with mineral surfaces; these may become overlaid with microbial or other organic products in zones of high microbial activity such as the rhizosphere. At present there is inadequate evidence on this question and vesicular arbuscular

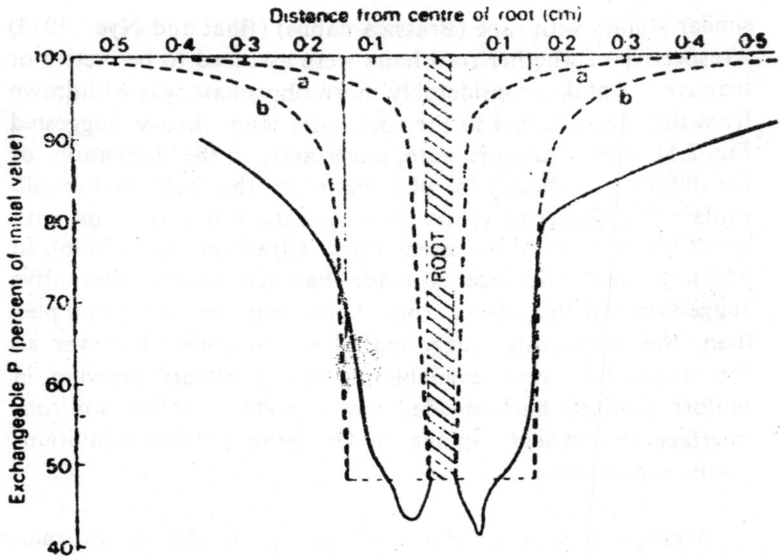

Fig. 2.5: Concentration profiles of phosphate in soil adjacent to a root of rape (Brassica napus) after eight days (solid line) compared with those predicted on the alternative basis that root hairs were inactive (dotted line a) or that uniform depletion occurred within the root hair cylinder (dotted line b). The diameter of the root hair cylinder is indicated by light shading

mycorrhizas are the only microbial components of the root/soil interface which have been proved to modify nutrient movement. They can convey phosphate to roots considerably more rapidly than could result from diffusion through the bulk soil or can occur when they are absent.

The need for a fuller understanding of ionic relationships at the soil/root interface is further emphasized by a comparison of relationships between the external concentration of nutrients and absorption when plants are grown in flowing culture solutions, in which the ionic concentration is maintained constant (*e.g.*, Clement *et al.*, 1974), with those observed in solid media. Wild *et al.*, (1974) have shown that when plants of several species were grown in sand culture (in which ionic interactions with the solid phase should be less than that in soil) it was necessary to maintain the concentration of potassium at more than a hundred times that in flowing cultures to achieve maximum uptake and growth, even though fresh nutrient solution, in a volume equal to that which the sand would hold against drainage, was added daily to the sand cultures. There is at paesent no evidence to establish that this difference can be explained solely in terms of the rate at which potassium diffused towards roots in the sand cultures.

THE ROLE OF ROOT HAIRS

Unlike some components of the root/soil interface, root hairs are clearly visible and their possible role in the absorption of water and nutrients has been frequently suggested. But there is considerable uncertainty as to their importance from either view point. Newman (1974) has recently reviewed evidence on their role in the absorption of water. He has drawn attention to the slender basis for the common assumption that root hairs enter water-filled pores in soil and also gives reasons for concluding that laboratory demonstrations of water uptake by root hairs provide little guidance on their effectiveness in soil. Theoretical considerations lead him to the view that the conductivity of root hairs, at the density of 100 cm^{-1} root length, may be only one-twentieth of the conductivity of the soil which surrounds the root cylinder even when the matric potential is—15 bar. None the less, the facts that the development of a root/soil 'gap' can sometimes restrict the uptake of water and that root hairs may anchor parts of roots to the soil suggest that it would be imprudent to dismiss them as unimportant in all circumstances.

Several types of evidence have been advanced to suggest that root hairs can accelerate the uptake of nutrients. Lewis and Quirk (1967) demonstrated that phosphate labelled with ^{32}P was rapidly removed by roots from within the root hair zone. Studies of the diffusion of both phosphate and potassium, to which reference has already been made, sometimes indicate that these ions enter roots at a greater rate than could be explained by their diffusion in the bulk soil to the surface of the root itself, but that diffusion could account for their arrival in the cylinder of the soil explored by root hairs. However, these observations do not by themselves establish that root hairs play a major role in absorption; the same results would be obtained if diffusion were considerably more rapid in the soil explored by rhizosphere organisms. The lack of conclusive evidence on this question has already been noted.

A different approach to the role of root hairs in nutrient uptake was adopted by Barley and Rovira (1970) who inserted roots into channels punched in samples of a sub-soil which had been compacted to different extents; over relatively short periods the absorption of phosphate was greater when the pore diameter permitted the lateral growth of root hairs. The conclusions which these observations warrant with respect to the function of root hairs depend on: first, whether variations in the compaction of the soil affected root function in ways other than root hair development and, second, whether the experimental conditions realistically simulated normal situations; the channels were of slightly smaller diameter than the roots and the extent if any of surface smearing in making the channels is not reported.

A further and more tenuous argument which sometimes obtrudes into discussion on the function of root hairs is that their presence implies some function, absorption being the obvious one since roots are absorbing organs. With increasing knowledge the plausibility of this approach has decreased. Other functions for root hairs have been identified—they anchor roots as they penetrate the soil and they can be secretory organs. Moreover, the fact that they can be rapidly colonized by micro-organisms could suggest that they may be significant mainly as sources of substrate for the rhizosphere. Furthermore, the extent to which, in normal field situations, root hairs do increase root/soil contact seems uncertain. They can readily be observed ramifying in the finely divided soil (Drew and Nye,

1969) but in field conditions they are often most conspicuous on roots growing in voids of appreciable size, contact thus seeming to be with the soil atmosphere at least as much as with the solid phase.

The difficulty of obtaining unequivocal evidence on the quantitative importance of root hairs as absorbing organs under natural conditions is obvious. In the writer's view no definite conclusion is now justified. Their effect may well vary widely depending upon soil conditions; the possibilities cannot be discounted at the present time that they may absorb and translocate nutrients to an important extent, that nutrients migrate more rapidly along their outer surface than through soil, or that they affect nutrient movement through their influence on the rhizosphere flora. The recent conclusions of Brewster *et al.*, (1976) are here of considerable interest. Detailed studies of the diffusion of phosphate through soil towards roots led them to conclude that the assumption that root hairs could be regarded as simply increasing the effective radius of roots was inadequate; the possibility that exudation from root hairs might exert a significant effect was referred to. Definitive information is lacking but these physico-chemical conclusions strengthen the view that uncertainties with regard to the role of root hairs are unlikely to be satisfactorily resolved until the combined techniques of plant physiology and microbiology, as well as the physical chemistry of the soil, are brought to bear on this problem.

ROOTS AND SOIL

The words 'plant growth' not 'root growth' are intentionally used in the title of this section. Equally in the study of the ecology of natural vegetation and of agriculture, root systems deserve consideration primarily because of their contribution to the growth of the plant as a whole. Their intrinsic characteristics are of secondary interest. Phrases such as 'ideal conditions for root growth' are not uncommonly used in relation to the cultivation of the land but ideal soil conditions for plant growth are not necessarily those which lead to the maximum proliferation of roots. They are those which permit the shoot system to photosynthesize and develop at the maximum rate which the aerial environment and genetic factors permit when the supply of water, nutrients and growth substances from roots impose no limitation.

If the rooting medium is well aerated and constantly supplied with adequate water and nutrients, other conditions being favourable, very restricted root systems can support considerable shoot growth. This principle is well illustrated by some systems of intensive glass-house culture in which there can be advantage in confining root systems to relatively small volumes of soil, though the reasons for this are uncertain (Cooper and Hurd, 1968; Cooper, 1971). With field crops the situation can be very different. Unfavourable soil conditions, which restrict the performance of roots are a common reason why the yield of crops is usually much below the potential maximum which their genetic characteristics, the incident solar radiation and the temperature would allow. A major objective in the study of root/soil relationships is thus to identify the restraints or stresses in the soil environment which can restrict the growth of plants. These restraints being identified, the aim of agriculture and horticulture is to mitigate them.

The stresses which roots can experience in soil fail into three broad, though often interrelated, groups: chemical stresses caused by the shortage of nutrients, their unbalanced supply or by the presence of toxic substances, physical stresses caused by an inadequate water supply, by the mechanical impedance to root penetration, by anaerobic conditions or by unfavourable temperature, and biological stresses caused by the flora or fauna of the soil. The most conspicuous biological stresses are usually due to plant pests and diseases, subjects which in general lie outside the scope of this book; other biological aspects are most conveniently considered in relation to the physical or chemical conditions which they create or which favour their occurrence.

Chemical Stresses

What is sometimes described as the 'modern period' in the study of soil/plant relations started in the early part of the nineteenth century (for review see E.W. Russell, 1973). It was soon recognized that crop yields were frequently limited by an inadequate supply of nutrients in the soil and agricultural chemists found means to mitigate this. The development of artificial fertilizers has been one of the principal causes of increasing crop yields and, without doubt, the most outstanding contribution of soil science to the productivity of agriculture. Progress was already considerable many decades before detailed studies began on the physiological processes which control the growth and function of root systems. The early agricultural chemists had realized that simple chemical analyses of plants and of the bulk soil, together with observations of the effects of added nutrients on crop yields, were often a successful and practical method for identifying nutritional requirements. These procedures have been progressively refined and although many aspects of soil chemistry are still subjects for active research—particularly methods for using fertilizers most effectively—it can be fairly claimed that 'fertilizers eliminate one natural limitation to crop production—the supply of plant nutrients from soil' (Cooke, 1970).

Physical Stresses

No statement comparable to Cooke's can be made about limitations imposed by soil physical factors. Sometimes they can be alleviated but often not eliminated by drainage, irrigation or cultivation; they are the major determinants of land use. Table 2.2 shows the land use capability classes, based largely on the physical characteristics of the soil, which it has been found convenient to recognize in the United Kingdom (Bibby and Mackney, 1962); the classification is similar to those introduced earlier in the United States. Any such classification is inescapably arbitrary and soils assigned to relatively low capability classes, for example Class 3, may sometimes yield exceptional

TABLE 2.2

Land Use Capability Classes in Great Britain

Class	*Capability*	*Characteristics*
1.	Very minor or no limitation to use .	Usually well drained, sandy to silty loams; good available water holding capacity; level or gently sloping well supplied with nutrients or responsive to fertilizers; favourable climate.
2.	Minor limitations which reduce choice of crop and interfere with cultivation or harvesting .	One or more of the following limitations—moderate or imperfect drainage; less than ideal rooting depth; unstable physical structure; moderate slopes; risk of erosion; climate not optimal.
3.	Moderate limitations which restrict choice of crops and/ or careful management necessary. Mainly suitable for grass, cereals and forage crops.	One or more of the following limitations imperfect or poor drainage; restricted rooting depth; poor structure; sloping ground and possible erosion; moderately unfavourable or severe climate.
4.	Moderately severe limitations to choice of crops and/ or requires very careful management. Grass usually the main crop.	One or more of the following limitations—poor drainage which is difficult to remedy; occasional flood damage; shallow and/or strong soils moderately steep with risk of erosion; moderately severe climates.
5.	Such severe limitations that pastures, forestry or recreation are only uses.	One or more of the following limitations which it is impracticable to correct—bad drainage; flood damage; severe risk of erosion; severe climate.
6.	Very severe limitations which restrict use to rough grazing, forestry or recreation. Pasture improvement impracticable.	Characteristics as in Class 5 but more severe; in addition soil may be shallow with stones or boulders.
7.	Land with extremely severe limitations which cannot be rectified and at best produce poor rough graving for a short period each year.	Examples are rocky slopes, sand dunes.

crops. But this does not dispute the general usefulness of the system.

Only a fraction of the land available for agriculture in many countries is of Class 1 and 2; these are generally the most intensively cropped soil. The expansion of farming to meet the needs of the increasing world population therefore often depends largely on improving the fertility of land which is of Class 3 or lower. For this reason alone, the development of methods to offset physical restraint in the soil deserves increasing emphasis in agricultural research. Other reasons point to the same need. In Western countries the use of heavier machinery can create the risk that soil structure will be damaged by compaction and the elaborate tillage, on which farmers formerly relied to maintain fertility, is often no longer practicable because of the cost and scarcity of labour. Different constraints have been experienced in other environments and the need to prevent soil erosion, when cropping is intensified, has sometimes been the major problem. These developments have led to considerable innovation in tillage systems.

Modern constraints have thus tended both to extend arable farming to soils in which physical problems are more likely to occur and at the same time to make it less easy to employ some of the traditional methods for offsetting any deleterious effects. It will be shown in that the new and simpler methods of cultivation can sometimes provide important opportunities for maintaining the fertility of the soil, but if the maximum advantage is to be taken of them, without undue risk, the response of root systems to their physical environment must be more fully understood. This is from several points of view a more difficult subject than the study of field problems of crop nutrition and one which until recently usually received less attention.

The foregoing reasons make the effect of the physical environment an appropriate central theme in a discussion of root/soil relationships at the present time. Thus, the following topics have been selected for fuller discussion in the chapters which follow:

— the manner in which mechanical restraint caused by the solid phase of the soil affects root growth,

— the effects of anaerobic soil conditions,

— the manner in which roots establish contact with the soil, and

— the effects of cultivating the soil on crop growth, a subject which enables much of the foregoing discussion to be set in a practical context.

In the remainder of this chapter some of the physical and chemical characteristics of the soil, which are of major concern from these viewpoints, are considered briefly.

THE SOIL AS A MEDIUM FOR ROOT GROWTH

Between 43 and 60 per cent of the total volume of the soil commonly consists of its solid fabric—mineral matter derived from the parent rocks, and sometimes calcium carbonate from pre-existing organisms, together with varying quantities of relatively stable organic matter which is usually most abundant in the surface layers. This solid fabric is penetrated by irregularly shaped pores, or voids, filled either with air or water. On them the life of the soil depends. Oxygen enters the soil mainly through air-filled pores and the movement and retention of water in soil is also largely dependent on its porosity. Moreover, the presence of pores or voids which roots can enter freely, or can expand by resisting only small pressures, is an essential requirement for their growth. Thus, in a discussion of the soil as a medium for root growth some of the questions of most immediate interest are the pattern of pores and voids and the manner in which the supply of oxygen, water and nutreints is maintained in the soil solution adjacent to roots. The composition of the solid phase itself is referred to here only insofae as it immediately affects these properties. The reader is referred to standard works on soil science for a more comprehensive treatment, *e.g.*, Baver *et al.* (1972), E.W. Russell (1973) and Brady (1974) as well as texts on individual aspects which are cited in the pages which follow.

Pore Space and Voids

Many of the individual particles which form the solid fabric of the soil are normally bound together to greater or less extent to form aggregates of varying size and shape, ranging from small crumbs to clods. The size of pores both within and between the aggregates can differ much, both between soils, depending largely on the composition of the solid phase, and in any one soil with depth.

The clay content of soil is an important determinant of pore size (Table 2.3); in clay soils the pores larger than 30-60 μm are relatively few but small ones, especially those less than 0.2 μm are more abundant and the total pore volume can be somewhat greater than that in loams.

On this 'fine structure' of the soil there is normally superimposed a 'macrostructure' clearly visible to the naked eye. Fissures and channels often occur through the solid matrix. They arise from many causes. If soils shrink on drying cracks develop and when the soil expands again, with the addition of water, planes of weakness may remain. The decay of roots and the soil fauna, for example earthworms can also create channels.

Retention of water in pores and its implications for soil structure. The size of pores determines the potential of water within them and consequently the suction necessary to withdraw it. Table 2.3 shows that the suction necessary to withdraw water from pores is considerably greater if their diameter is small. Thus, if increasing suction is applied to a soil, the distribution of pores by size can be inferred from the quantity of water which is withdrawn at different suctions; this relationship between the suction applied and the quantity of water held in soil is commonly expressed graphically as the 'moisture characteristic curve'.

TABLE 2.3

Pore Space Distribution in Some Contrasting Soils

	Soil type			*Approximate suction to withdram water from pores (bar*)*
	Light sandy-loam	*Medium loam*	*Clay*	
Percentrge clay	*>10*	*20-40*	*>40*	
Pore size (μm)	*Per cent of total soil volume*			
>300	>1	5	2	0.01
96-300	3	1	0.5	0.03
60-96	5	1	0.5	0.05
30-60	7	4	1	0.1
6-30	15	9	7	0.5
0.2-6	<1	20	22	15.0
0.2	15	10	22	>15.0
Total pore space	45	50	55	

*The potential with which water is held is numerically equal to the suction but with a negative sign.

The retention of water against gravitational forces in the larger pores is due primarily to capillary forces, but in addition, water interacts with the lattice structure of clays. The mechanism by which water is retained has an import influence on soil properties. If water is withdrawn from stable pores in which it has been held by capillary forces, air replaces the water, the volume of the soil being little affected. However, if water has been retained by interaction with clay its withdrawal causes clay platelets to come closer together with the result that the clay shrinks. Thus, if the pore space in an appreciable mass of soil consisted predominantly of fine pores (*e.g.*, less than 0.2 μm) between clay lattices, shrinking on drying would cause considerable internal tensions to develop; these are eventually relieved by the cracking of the soil, sometimes to a considerable death, leaving massive blocks of harsh structure which can be intractable to cultivate. If on the other hand the soil is penetrated by a sufficient number of large pores in which water had baen retained by capillary forces, the soil shrinkage will not only be less but it will cause the soil to fragment along the planes of weakness created by the air-filled pores, thus yielding a friable crumb structure.

The examples considered above represent two extremes; those found in practice are often intermediate and, as it will be shown later, the organic matter in soil as well as dissolved substances can also influence soil structure. None the less the larger pores in which water is retained by capillary forces always have an important effect, particularly for the management of heavy clay soils. When the soil is compressed the volume occupied by large pores is particutarly reduced and as the 'strength' of soils, that is to say their resistance to deformation, decreases with rising water content the compression of heavy wet soils by natural process or cultivation can impair soil structure through the elimination of the larger pores.

Pore space, soil water and aeration. The pores and voids in soil can, on a simple analysis, be considered in three broad categories, which together with fissures and channels in the macrostructure control aeration the infiltration of water and drainage.

Pores which drain freely under gravity so that they are air-filled when the soil is at, or close to, field capacity (water potential c.—0.1 bar). The minimum size of pores in this class is c, 30-60 μm. Because the diffusion coefficient of oxygen in the gas phase is some 10,000 times greater than that in solution the aeration of the soil when it is close to field capacity, as well as drainage, depends largely on these pores.

Pores which hold water against gravity but in which the water potential is sufficiently high for absorbing roots to withdraw it that is to say the water potential is, above—1·5 bar (*i.e.*, the wilting point. The minimum size of such pores is c. 0.2 μm. The quantity of pores between c. 0.2 and 60 μm is a major factor which determines the reserves of water which are available to plants. When water is withdrawn from these pores they become filled with air, thus contributing to the aeration of the soil.

Fine pores which hold water at potentials less than—15 bar so that is inaccessible to plants.

Variation in pore size distribution which commonly occurs at different depths from the surface of field soils can, however, modify the relationship between pore size and the downward transfer of water. Its movement depends on the existence of a gradient of decreasing potential caused either by gravitational or matrix forces. Since the potential of water is lower in small pores than in large ones (see Table 2.3) it will not pass as readily from small pores, in which it is held against gravity, to larger ones, as along pores of a relatively constant average size. Accordingly if a layer of soil with a relatively large pore diameter underlines one of smaller pore diameter the downward movement of water may be much retarded even though there is a considerable air-filled pore space in the underlying soil. In this way quite narrow zones in which pore size is small can much restrict the downward movement of water and the ingress of air. This effect can be brought about, for example, by the 'smearing' of the soil when ploughed under wet conditions or by the deflocculation or 'slaking' of the surface soil in wet conditions.

The processes which control the movement of water in the soil are considerably more complex than the foregoing paragraphs suggest but detailed discussion of the physics of water movement lies outside the scope of this book; Childs (1969) and Slatyer (1967) are among the standard sources of fuller information.

Pore space in relation to root penetration. It will be that the rate at which roots elongate is much reduced if they have to exert quite small pressures to enlarge pores smaller than themselves. Thus, the existence of a sufficient number of continuous pores which roots can enter freely is an important requirement for their growth. The majority of roots (as opposed to root hairs) exceed 60 μm in diameter and are usually considerably larger, for example,axes and first order laterals of cereals may range in size from c. 150-750 μm; the pores they can enter are therefore air-filled when the soil is below field capacity. Since living roots represent but a small fraction of the total soil volume—usually less than five per cent—good root development does not require more than a small fraction of the total soil volume to consist of continuous pores equal in diameter to the large root members.

The characteristics of pore space in soil, which are necessary for the ready penetration of roots, are thus not identical with those which determine the movement of water or aeration.

Stability of the Pore Regime

Next to the size and frequency of pores their stability is their most important characteristic, but no soil is really stable in the strict sense. A soil which displays approximately constant characteristics over long periods and is therefore usually described as stable, is one in which processes which destroy its structure are balanced by those which create it.

The composition of the mineral fabric of the soil has an important influence on its stability. One example of this has already been provided in earlier discussion; in soils with a high clay content, and consequently a relatively small volume of pores of appreciable size compression under wet conditions may have particularly deleterious consequences.

The stability of the soil is also influenced by many other characteristics of the mineral phase which are extensively discussed in the general literature of soil science. A relatively high ratio of exchangeable calcium to sodium can decrease the extent to which soil crumbs deflocculate or slake under wet conditions. In addition to disrupting the pore structure, slaking can release collodial material which is subsequently carried down and deposited at greater depth, thus blocking voids between structural units. Moderately light soils, in which an appreciable quantity of silt and sand of various sizes favours the development of relatively large pores, are often more stable.

In a discussion of root function and the soil it is appropriate to consider in more detail the ways in which plants themselves contribute to the maintenance of a stable pore regime. The importance of the effects of roots is evident from the sequence of events when changes in climate, or geomorphological processes, permit plant colonization to begin on purely mineral substrates, for example, sand dunes or moraines deposited by glaciers. Until plant growth commences there is no soil, as that word is commonly understood, merely mineral particles of various sizes which usually cohere little one to another except when substances carried down in solution are precipitated on them. The creation of a stable pore structure by the action of roots and the fact that in their absence the soil can be rapidly degraded is well illustrated in agriculture by the improvement of soil structure which can occur when land is maintained as permanent grassland as opposed to being cultivated annually so that it supports growing plants for only a part of each year.

The contrasting characteristics of soil derived from the same mineral material to which these different systems of management lead have been well illustrated by Low (1972) and Low and Stuart (1974). They compared adjacent fields of pedologically similar calcareous clay which had either supported grassland or been under continuous arable cultivation for at least a century. Under permanent grassland the soil was penetrated by pores or planes of weakness through which roots ramified, but prolonged arable cultivation caused a much more compacted structure. The pore structure of the two soils is shown in Table 2.4 and also that an another field which has been under continuous grass until it was ploughed a few years previously.

TABLE 3.4

Effect of Management on Pore Space and Density of Chalky Boulder Clay soil (Hanslope Series)

Description of field	*Percentage of total soil volume*			*Apparent density*
	Total pore space	*Occupied* by water*	*Occupied* by air*	*g cm^{-3}*
Old arable	41.8	36.8	5.0	1.47
Old grassland unploughed	55.6	35.3	20.2	1.08
Old grassland ploughed four years enrlier	55.3	47.5	7.8	1.09

*Measurements were made when the soil was close to field capacity.

The 'old' grassland soil had a lower bulk density than the 'old' arable soil and larger pore space; its considerably greater air-filled volume reflected the presence of much more numerous pores or channels which were sufficiently large to drain freely. However, four years of arable cultivation reduced the volume of air-filled pores to little more than that in the old arable field.

These differences in pore size correlate with differences in the stability of soil aggregates which was greatest on the old grassland soil; a marked decrease occurred after ploughing for even only one year (Fig. 3.6). Detailed examination with the electron microscope (Low and Stuart, 1974) showed that an adherent, presumably organic, matrix bound the particles surrounding pores in the old grassland soil; it was absent in the old arable land. The difference in aggregate stability was, however, not related to the total organic matter content of the soil; the large decrease in stability caused by ploughing for a single year (Fig. 2.6) was accompanied by only a small reduction in total organic matter.

Many other observations point to the same conclusion; the stability of the soil appears to be determined not so much by its

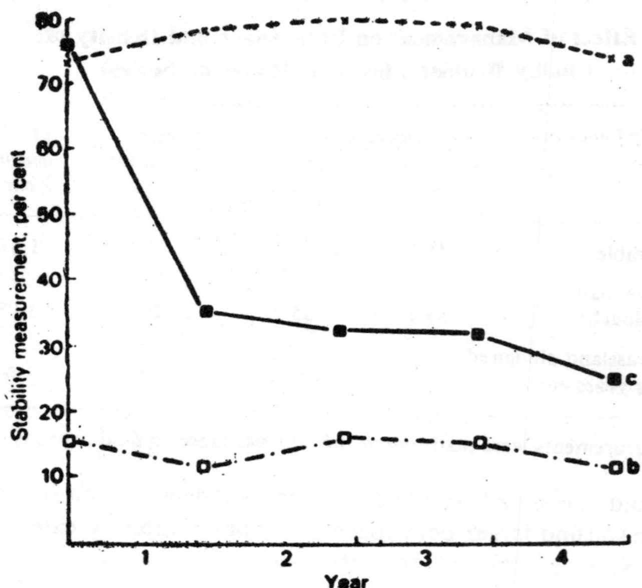

Fig. 3.6: Influence of cultivation on the water stability of air-dry aggregates 3-5 mm in diameter in the upper 15 cm of the calcareous clay soil (a) Permanent grassland for c. 100 years. (b) Continuous arable for c. 100 years. (c) Permanent grassland ploughed at the beginning of the period of observation and then cultivated annually Stability was measured in the autumn of each year

total content of organic matter but by some components of it which can readily be destroyed. Polysaccharides of bacterial origin are believed to be involved. Detailed information is scanty (see E.W. Russell, 1973 for review) but the beneficial effects of growing roots on soil structure encourages the view that rhizosphere organisms which utilize root exudates may be an important source of these substances. But they are not the only ones. Members of the soil fauna for example earthworms, can also stabilize the soil. The incorporation of organic matter—dung or compost—or the use of artificial soil conditioners of which polyvinyl alcohol is one of the best known, can have a comparable effect.

While Table 3.3 appear generally representative of the manner in which soil structure can be modified by root growth it must be emphasized that the magnitude of the effect can vary greatly. Apart from the nature of the mineral substrates climate can have a considerable influence; high soil temperatures, *e.g.*, greater than 30°—40 °C can accelerate the breakdown of soil organic matter causing structural deterioration. Low temperatures can retard the creation of new organic matter due to the reduced rate of plant growth. However, if the temperature falls below 0°C soil structure can benefit though the expansion of pores caused by the freezing of water.

Nutrients in the Soil Solution

The role of the soil solution as a source of nutrients for plants and the manner in which the concentration of the solution in the near vicinity of roots is maintained are the subjects of most immediate relevance in relation to plant nutrition. They alone are discussed here; the chemical and biological factors which determine the supply of 'available' nutrients in the bulk soil are extensively discussed in the literature of soil science.

Importance of the soil solution. The solution phase in the soil adjacent to roots is widely accepted as the immediate source of ions which enter roots except when mycorrhizas convey nutrients from sites in soil some distance from roots. An alternative suggestion has, however, been made that ions which are held by exchange or other processes in the solid phase (and this includes all cations, as well as some anions) may enter plants directly from soil surfaces adjacent to roots by 'contact exchange' (Jenny and Overstreet, 1939); this theory envisages, for example, that hydrogen ions released from plant roots could exchange with potassium ions held in exchangeable from on the surface of clay minerals. There is no reason to doubt that when hydrogen ions are released by roots they may exchange with cations in the solid phase, sometimes at sites in very close proximity, so that their pathways through the soil solution are very short, but there appear no grounds for assuming that the solution phase is entirely bypassed. Modern concepts on the 'apparent free space, within roots and the interposition of mucigel between roots and the soil are more readily compatible with the view that ions pass into them in the solution phase. Moreover results of studies of the passage of ions through the soil, appear to be adequately explained in terms of their arrival at or near the root surfare either by mass flow or diffusion through the soil solution. The concept of 'cation exchange' will therefore not be considered further. At any instant in time the concentration of a nutrient in the solution immediately adjacent to a root appears to be the best measure of availability for absorption though, many factors within the plant and also the concentration of other ions in the solution phase may influence the actual rate at which it is absorbed.

Replenishment of the soil solution. As living roots occupy only a small fraction of the total soil volume, the rate at which nutrients diffuse through the soil towards roots can have an important influence on the rates at which they are absorbed by plants, Some representative values for different ions diffusing through the soil are shown in Table 3.5. Both chloride and nitrate move at about a quarter of that in free solution; they interact little with the solid phase in the soil and their slower movement than in free solution is mainly because of the tortuosity of diffusion pathways through the solid matrix and their smaller cross-sectional area. In contrast potassium, and still more phosphate, diffuse through the soil at only a small fraction of the rate observed in solution. Both interact with the solid phase. Potassium, like other cations, is held by exchange on the negative charges which predominate on clay and other surfaces while phosphate can be sorbed by oxides of hydroxides of di- or tri-valent cations or in other ways. The rate of diffusion of ions which interact with the solid phase in these ways can vary widely depending on their concentration, on the composition of the solid phase, and on the moisture content of the soil.

TABLE 3.5

Ciffusion Coefficients for Some Ions in Solution and in Moist Soils at 20-25°C

	Diffusion coefficient ($cm^2\ s^{-1} \times 10^5$)	
	Free solution	Soil
Nitrate	1.92	0.5
Chloride	2.03	0.5
Potassium (or Rubidium)	1.98	0.01-0.24
Phosphate	0.89	0.0005- 0.001

Nutrient ions which can diffuse freely in the soil reach plants primarily by mass flow in the water, which moves towards roots to satisfy the requirements of transpiration; it seems likely that this is true also of non,polar molecules which plants can absorb, for example urea or mineral chelates. These substances are thus constantly replenished in the soil solution adjacent to roots. Indeed ions which move readily by mass flow but enter plants to only a limited extent, for example sulphate, may accumulate at the exterior of the root (Wray and Tinker, 1969). In some soils calcium behaves similarly (Barber and Ozanne, 1970).

In contrast the slow diffusion of phosphate and potassium can cause them to become depleted in the solution phase near, roots. The limitations on absorption which are imposed by the slow movement of phosphate to roots has been vividly illustrated by Lewis and Quirk (1967) who concluded that over 94 per cent of the phosphate absorbed by roots during four days had come from soil within 0.1 mm of it. Detailed physico-chemical studies.

On a simple analysis it would appear that when the movement of ions towards roots is limited by diffusion the amount absorbed by a root system will vary with its surface area and total length since an increase in either of these dimensions will reduce the mean diffusion pathway through the soil; the rate at which roots elongate could also exert an influence through bringing them in contact with previously undepleted soil.

METHODS FOR STUDYING THE DISTRIBUTION AND FUNCTION OF ROOTS IN THE SOIL

Many methods have been described for studying the distribution of root systems in the soil. They contrast not only in the type of information they provide but also in the complexity of equipment which is required and effort which must be expended to obtain representative results. The object of these notes is to assist in the selection of the most appropriate procedure in different circumstances. The references cited have been chosen as being among the most recent which describe the procedures in sufficient detail to provide a basis for judging their suitability. It is usually important to relate observations on root distribution to measurements of soil physical or chemical factors. Methods by which these latter measurements can be made are, however, not reviewed as they are extensively covered in the general literature.

The Distribution of Root Systems

(a) Visual inspection. The appropriate first stage in defining questions for detailed study is usually visual inspection by digging out clods if roots are shallow or by examining the sides of profile pits. Apart from giving qualitative information on possible restraints in root growth (*e.g.*, evidence of mechanical impedance) the inspection can give some indication of the scale on which sampling will be necessary in subsequent quantitative studies. If root axes are few in number or occur mainly in relatively infrequent voids between massive structural units, more extensive sampling of the soil may be necessary that if roots are numerous and distributed relatively uniformly.

(b) Separation of intact root systems from soil. Weaver (1926) described a laborious procedure for washing intact root systems free of soil and recording their position in the soil profile. A number of more recent workers have used 'pin boards' for this purpose (de Roo, 1969); this method involves exposing a smooth vertical face of soil, against which a boards perforated with holes c. 10 cm apart, vertically and horizontally, is placed; steel 'pins' or rods are then driven through the holes into the soil to a sufficient depth to pass through the root system. When the soil is subsequently washed away the pins support the roots in approximately their original position.

While providing a general description of root from this method is likely to be less suitable than the method described below for quantitative studies; not only are larger errors likely from the causes noted in the discussion of that method but it is less easy to obtain representative information applicable to a plant population.

(c) Separation of rools from soil cores. Soil cores are obtained by diving hollow tubes usually 5 cm or more in diameter into the soil and subsequently removing them; simple mechanical devices such as those described by Welbank and Williams (1968) and Ellis and Barnes (1971) can facilitate this work. The soil cores are divided into appropriate sections and the roots washed free of soil (Welbank *et al.*, 1974). Root lengths in different horizons can be estimated by the 'intercept' method of Newman (1966) with simple mechanical equipment (Rowse and Phillips, 1974); the dry weight of the root can also be determined.

The main sources of error in this method are:

1. incomplete separation of roots from the soil and the loss of root fragments.
2. the difficuly, or impossibility, of distinguishing living roots which have lost their original white colour, from those which have recently died, or from roots of similar form which belong to other species.

Errors in separating roots can be reduced by soaking the soil in solutions of sodium pyrophosphate (10 g. l^{-1}). To obtain representative information on field experiments, it is necessary examine numerous cores; the presence of stones may make coring impracticable.

(d) Observation of the distribution of roots down the soil profile. In many investigations the depth which root systems attain is a question of particular interest. In the absence of complications due to weeds or stones this can be studied relatively simply by takihg soil cores (as in method c), cutting horizontal faces across them and counting roots after a few mm of soil have been washed away. To supplement this information, relationships can be established between root numbers on the horizontal faces and their length or weight per unit volume of soil by using the more labourious procedure of method (c) on some cores.

The magnitude of the errors associated with this procedure depend both on the number of cores examined and on the heterogeneity of the soil. It can be the most practicable way to obtain representative information in large-scale field experiments (Drew *et al.*, 1977); the methods described below are likely to be suitable only for intensive work on a limited scale.

(e) Estimation of the proportion of roots in different zones of soil by tracer methods. 1. Destructive sampling. These depend on injecting into plant shoots radioactive tracers which become sufficicntly uniformly distributed through the root system for the quantity of tracer present per unit volume of soil to reflect the volume (or weight) of living root which is present. The isotopes, ^{32}P (as H_2PO_4) (Racz *et al.*, 1964), ^{86}Rb (Russell and Ellis, 1968; Ellis and Barnes, 1973) and ^{42}K can satisfy this requirement. Their respective half-lives are 14.3 and 18.7 days and 12.5 hours. In contrast to ^{32}P, both ^{86}Rb and ^{42}K emit gamma radiation which is of sufficient energy for them to be accurately measured with suitable scintillation counters in relatively large volumes of soil (c 2 kg), thus giving representative information if the samples are compounded of appropriate sections of many cores obtained in the manner described in (c). Because of the low penetrating power of the beta radiation of ^{32}P this isotope can be measured only in very small samples. The assay of large numbers of samples, or the chemical separation of the tracer from the soil is therefore likely to be necessar to provide representative information. While it does not suffer from this disadvantage ^{42}K is inconvenient because of its short half-life except in the special application referred to below. Thus for field studies ^{86}Rb would be the obvious choice.

The advantages of the ^{86}Rb procedure are that sources of error noted with method (c) are avoided. Its main disadvantages or limitations are:

(a) elaborate equipment is necessary for the measurement of radio-activity in numerous samples are to be analyzed (Lay, 1973).

(b) the method is valid only if the tracer becomes uniformly distributed per unit volume of tissue throughout the different zones of the root system it is desired to compare (for example, perhaps the root present in a 2 cm cube of soil). It is known that this requirement is satisfied by some gramineaceous root systems and that loss either to the soil or dead roots is negligible (Russell and Ellis, 1968), but it caanot be assumed that the situation is similar in root systems of contrasting type and

(c) as applied hitherto, the method estimates the fraction of the root system in different zones of soil, not the actual weights or volumes therein. Moreover, it cannot be used in the surface soil, in which the bases of shoots contribute much of the radio-activity.

2. **Non-destructive sampling.** In some types of detailed study successive observations of the distribution of the same root systems may be desirable. This can be achieved by using ^{42}K. Measurements of gamma activity are made at different depths in soil by lowering a scintillation counter down an aluminium

tube of the type used with neutron moisture meters (Mercer *et al.*, 1975); after the tracer has decayed (*e.g.*, 1-2 weeks) the operation can be repeated. The main complexities of this method, which is in an early stage of development, arise in part from the limited period during which observations must be made; moreover a computer programme is necessary to calculate root distribution from the observed counting rates.

(f) Location of individual roots id detailed studies. Baldwin *et al.*, (1971) have described an autoradiographic technique to show the positions, on a plane in the soil, of the individual roots of plants to which radio-active tracers have been applied; double labelling could enable the roots of different plants to be distinguished. A still more elaborate method is to impregnate soil with resin and cut sections by methods used with minero-logical samples (Burges and Nicholas, 1961, Melhuish and Lang, 1968). If changes in the volume of both soil and roots can be prevented this method could be well suited for studying soil/root contact.

The Absorption of Nutrients and Water

Nutrients. The labelling of the soil at different depths in parallel plots with radioactive tracers provides an opportunity to compare the quantities of nutrients which are absorbed from them. This method has been mainly used to study the uptake of phosphate and calcium (Hall *et al.*, 1953; Newbould, 1969; Newbould *et al.*, 1971), ^{32}P being used as a tracer for phosphate and ^{89}Sr for calcium (this isotope can be assayed more readily than ^{45}Ca and has been shown to be an adequate tracer for calcium in the soil/plant system). The methods appear to be unsuitable for work with the majority of other nutrients because of the absence of suitable radioactive tracers; for example the half-lives of the isotope of potassium are too short and no isotopes of chemical homologues are suitable. Moreover there would be considerable difficulties in using stable isotopes in this way.

These tracer methods give valid information only if:

1. Allowance can be made for any variation in the extent to which the tracer equilibrates with the labile nutrients in the soil between different zones of injection.
2. The procedure used to inject the tracer does not cause any appreciable change in the quantity of labile nutrients present. This requirement is due to the fact that localized enrichment with nutrients can modify root form and nutrient uptake; carrier-free isotopes must thus be used.

Methods for meeting these requirements have been described (*e.g.*, Newbould and Taylor, 1964).

The use of fertilizers labelled with isotopes has also been suggested to determine the optimal method of placement in agricultural practice. It has been assumed that the efficiency of different methods of placement can be inferred from a comparison of the amounts of tracer absorbed by plants when labelled fertilizers are placed in different position. One advantage claimed for the method is that generalizations on the optimal method of placement can be made even when the nutrient supply in the soil is such that there is no response to the fertilizer (*e.g.*, IAEA, 1970a, b). However, the validity of this approach appears to be questionable, since:

1. In the study of the use of fertilizers, interest centres on the increase in the total nutrient content of the plant to which it leads, not on the uptake of fertilizer alone. This can be an important distinction because localized enrichment of nutrients can either reduce or enhance absorption from other zones of soil, the mechanism which bring about the former effect is evident from, while the latter can result when the increased growth of roots, caused by the enhanced nutrient supply, enables them to explore zones of soil which they would otherwise not reach.
2. When the placement of fertilizers increases plant growth the distribution of root systems can be modified.

> Accordingly the amount of a fertilizer absorbed from a placed fertilizer, which induces a response in growth, cannot necessarily be inferred from observations when there is no such response.

Water. The quantity of water which roots absorb from different zones of soil can be inferred from the manner in which soil water is depleted. The neutron moisture meter can be used for this purpose.

The measurements reflect both the uptake of water by roots from different depths and its movement through the soil caused by gradients in water potential which are created in part by the removal of water by roots. The marked interspecific differences evident in however indicate that the pattern of absorption by roots can be the major cause of variation in water content of the soil.

Artificial Systems

'Root laboratories' or 'rhizotrons.' The essential feature of these facilities is that the plants are grown adjacent to subterranean passages with transparent sides through which roots can be observed. Installations designed to study root systems of fruit trees (Rogers, 1969; Hilton and Khatamian, 1974) and herbaecous plants have been described (Glover, 1967; Taylor, 1969; Taylor *et al.*, 1970). The usefulness of the information which can be obtained when observations on root growth and soil physical conditions measurements are made conjointly in these installations is illustrated elsewhere in this book. The validity of the quantitative conclusions, which these observations justify, depends on root growth adjacent to the panels being similar to that throughout the rooting zone. There is evidence that this requirement can sometimes be satisfied when rhizotron compartments are carefully filled with uniformly mixed soil (Taylor *et al.*, 1970) but it is less likely that the results will be equally representative when transparent panels are placed against vertical faces cut in many undisturbed natural soils.

Lysimeters. The essential feature of lysimeters is that they are enclosed volumes of soil from which water lost by drainage can be collected, or changes in water content can be defermined by weighing. Provided that lysimeters are large enough to simulate reasonably the conditions to which it is intended to apply the results, they are a valuable facility for studying water and nutrient balances in the soil/plant system or for examining the effects of variations in water tables. Both for this reasons and because it can be possiblé to maintain more uniform soil conditions in lysimeters than in field plots they can provide convenient opportunitíes for employing some of the methods for studying the distribution and function of root systems which were discussed earlier in this section. General discussions of lysimeter design include those of Williamson *et al.*, (1969); Lal and Taylor (1969); Shalhevet *et al.*, (1969); Shalhevet and Zwerman (1962); Armijo *et al.*, (1972); Keller *et al.*, (1973).

4

Effect of Anaerobic Soil Conditions

The adjective anaerobic indicates the absence of free oxygen Under natural conditions the entire soil is never anaerobic as some oxygen always enters the surface layers and if the partial pressure of oxygen decreases below atmospheric, depletion is seldom uniform throughout an appreciable volume of soil. When anaerobiosis supervenes, it usually occurs first at localized sites.

The first part of this chapter is concerned with the factors which lead to these anaerobic soil conditions and with the consequent alteration of the soil as an environment for root growth. Subsequently, some aspects of the complex and imperfectly understood manner in which plants respond to anaerobic conditions are discussed. Agricultural implications are referred to in the final section.

THE DEVELOPMENT OF ANAEROBIC SOIL CONDITIONS

Gas Exchange Between the Soil and the Atmosphere

Anaerobic soil conditions occur only when the rate at which oxygen enters the soil from the atmosphere is less than that at which it is utilized in the respiratory process of plant roots, bacteria fungi or other organisms.

The air-filled porosity of the soil is its physical characteristic which has the greatest influence on gas exchange with the atmosphere; this is because oxygen diffuses in the gas phase some ten thousand times more rapidly than in solution. The saturation of the soil with water—waterlogging—is thus the most common cause of anaerobiosis but waterlogging, or flooding, does not necessarily have this effect. If the hydraulic conductivity of the soil is sufficiently high, and drainage is unimpeded, the movement of aerated surface water through the soil may provide sufficient oxygen; this, for example, happens in flooded water meadows especially in cool conditions when relatively little oxygen is being used in biological processes. The principal effects of the saturation of the soil with moving water, oxygen not being limiting, are the leaching of soluble nutrients, especially nitrate, and the proneness of the soil to compaction by traffic. Conditions for root and shoot growth can otherwise be highly favourable, in the same way as in a flowing solution culture.

There is no constant relationship between the air-filled pore space in a soil and the degree of anaerobiosis which can develop (Grable, 1966). Differences in the distribution of pores and their continuity can modify the transport of oxygen, both in gas and solution phases, to different zones in the soil. Moreover, variations in the rate at which oxygen is utilized can have a large effect. Table 4.1 from the work of Currie (1970) shows that the consumption of oxygen of a well drained field soil can change by a factor exceeding ten depending on the temperature. Table 4.1 also shows the influence of roots; the markedly higher respiration of the cropped soil reflects the respiration both of roots and of micro-organisms for which root exudates or dead roots provided substrates. E.W. Rsusell (1973) has reviewed numerous other estimates of the utilization of oxygen in field conditions: the majority fall within the range shown in Table 4.1 though there are suggestions of rather higher values in some tropical areas. He also points out that if a soil which contains

20 per cent by volume of air uses oxygen at the rate of 7 g per m^2 surface area per day the total oxygen contained in the soil air would be exhausted in about two days if its surface was completely sealed from the atmosphere. Thus, if oxygen is being used at the highest rate indicated in Table 4.1 the interruption of gas exchange for less than a day could lead to a marked depletion of oxygen.

TABLE 4.1

Oxygen Consumption in Winter and Summer by Soil Either hare of Crops or Under kale (Brassica Oleracea)

	July	*January*
Soil temperature at 30 cm	17°C	3°C
Oxygen consumed per m^2 ground surface (g d^{-1})		
Bare soil	11.6	0.7
Under kale	23.7	2.0

The absence of anaerobic conditions was established by the fact that the respiratory quotient was close to one throughout.

Localized Anaerobic Zones

When the rate at which oxygen diffuses into the soil starts to fall below that at which it is consumed, considerable differences in its concentration can occur between sites only a short distance apart. If the finer pores of the soil are filled with water, but the larger ones contain air the soil can, on a simplified model, be regarded as consisting of water-filled crumbs separated by air-filled pores. A number of investigators have considered relationships between the concentration of oxygen at the surface and in the centre of crumbs of different sizes under these conditions. Currie (1961) concluded that the maximum radius (r) of a crumb, to the centre of which oxygen would just reach, could be expressed by the equation.

$$r^2=6DC/M$$

where D is the diffusion coefficient of oxygen in the crumb which varies depending on the size and tortuosity of the water-filled pore space, C is the concentration of oxygen in the water on the outer surface of the crumb and M is the rate at which oxygen is utilized within it. It has been estimated that, depending on the rate of respiration, the conceatration of oxygen could fall to zero at the centre of water-filled crumbs of *c.* 0.1—1.0 cm in radius when their surfaces are bathed with water saturated with air (Greenwood, 1969: 1970).

The non-uniform distribution of organic substrates is an additional cause of variation in the concentration of oxygen within the soil, the rate at which oxygen is utilized being greatest when abundant substrates favour the proliferation of the microflora.

Fig. 4.1 attempts to illustrate in simplified form how anaerobic zones may develop when the water content of the soil increases. It is evident that:

1. the average concentration of oxygen in the gas phase of soils which are partially saturated with water is a most inadequate guide to the supply of oxygen available to roots,
2. aerobic and anaerobic processes can occur simultaneouly in adjacent zones of soil. This has two important consequences: first, that the growth of roots may be affected by diffusible products of anaerobiosis even though the roots themselves may be in aerated zones of soil and, second, that substances which arise only in completely anaerobic conditions may be found in soil in which an appreciable quantity of oxygen is still present; this situations is illustrated later in this chapter (*e.g.*, Fig. 4.3).

SOME EFFECTS OF ANAEROBIC CONDITIONS IN THE SOIL

The restriction of the supply of oxygen to roots is not the only

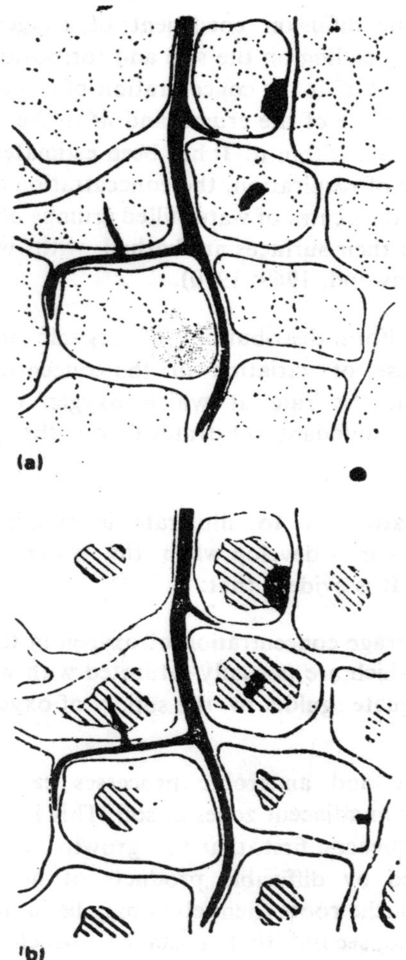

Fig. 4.1: Schematic representation of onset of anaerobic soil conditions, (a) Soil well aerated: Pores between aggregates are air-filled and there are smaller air-filled pores (dotted) in aggregates: a growing root and two zones with abundant organic substartes are shown (shaded). (b) Increasing soil water has displaced air in the fine pores within aggregates: Anaerobic zones (lightly shaded) are developing within aggregates, especially where substrates are abundant. The distance from air-filled pores to which the anaerobic zones may eventually extend is discussed on page 85

potential cause of injury to plants when anaerobic conditions develop in soil. Numerous and complex biological, chemical and physical changes occur; they are discussed in an extensive literature, *e.g.*, McLaren and Peterson (1967), Ponnamperuma (1972), Allison (1973), E.W. Russell (1973) and Skinner (1975). Attention is here limited to two aspects—the production by micro-organisms of toxic substances which may contribute to the initial injury of plants and losses of soluble nitrogen compounds from the soil which can militate againsl their survival or recovery. To set matters in context the generat nature of metabolic processes in soil is first considered.

Metabolic Pathways in Anaerobic Conditions

When free oxygen is absent many of the changes in the soil which can affect plant growth are due to the products of metabolism in obligate or facultative anaerobic micro-organisms. The majority of these are heterotrophic, depending on the oxidation of organic substrates for their energy, and a comparison of aerobic and anaerobic respiration indicates the general nature of some of the most important changes which can occur when anaerobiosis supervenes. Taking glucose as an example of a simple substrate, the two types of respiratory pathway can be summarized as follows:

$$6H_2O + 6CO_2 + 300\ 000\ \text{cal}^{*}$$

Aerobic
|
$+6O_2$
$C_6H_{12}O_6$
|
Anaerobic

$$2C_2H_5OH + 2CO_2 + 16\ 000\ \text{coal}^{*}$$

For the purpose of this discussion some of the most important similarities and differences between the anaerobic and aerobic processes are:

* The released of energy shown in the two equations is that which is biologically useful, being transferred through phosphorylation processes (Conn and Stumpt, 1966). The total release of energy as measured by calorimetry is considerably greater.

1. All respiration depends on the transfer of electrons from the substrate which is oxidized, to an acceptor which is reduced. In the aerobic process free oxygen is the electron acceptor combining with hydrogen ions to form water. When oxygen is absent a wide range of other reactions can occur. Combined oxygen in the substrate may be utilized, as in the above example; oxygen may come from the reduction of other substrates, such as nitrate or sulphate; cations of high valency, for example, trivalent iron or tetravalent managanese, may accept electrons and be reduced respectively to ferrous or manganous forms. The pathway of the electron transfer depends on the redox potential (see Table 4.2) and is influenced by pH and other factors. If the redox potential falls sufficiently, leading to extreme reducing conditions, hydrogen ions can accept electrons and give rise to molecular hydrogen.

TABLE 4.2

Oxidation-Reduction Potential at which Reactions Occur in Typical Soil Systems at 25° C and pH 5-7

Product of reduction	*Oxidation-reduction potential mV*	
H_2O (reduction of oxygen)	930	820
NO_2	530	420
MN^{2+}	640	410
Fe^{2+}	170	180
H_2S	70	220
C_2H_4	120	240
H_2	295	413

2. Whereas aerobic respiration can cause the complete oxidation of carbohydrates to carbon dicxide and water, anaerobic processes do not. They thus yield

much less energy and a wide range of partially reduced organic compounds result. These include alcohols (as in the examples quoted above) and numerous organic acids as well as other substances. Some of them may be decomposed giving rise to hydrocarbons and carbon dioxide, those which are volatile may escape rapidly from the soil but others may persist until they are metabolized in aerobic processes when a supply of oxygen is restored. Metabolic processes in anaerobic soils include those given in the following section.

Toxic Substances in the Soil

Many substance which can be produced by anaerobic metabolism can be injurious to plants. They can reach toxic concentrations if sufficient quantities of readily metabolized organic substrates are present. Their effects can be particularly conspicuous when massive quantities of slurry produced by intensive animal production are applied to the soil (Elliott and McCalla, 1972; Burford, 1976) but the incorporation of plant debris can be sufficient to cause significant effects, particularly when the temperature is favourable for rapid anaerobic decomposition.

The micro-biological products found in anaerobic soils have been extensively reviewed, for example, by McLaren and Skujins (1971) and in more general texts on soil science, *e.g.*, E.W. Russell (1973). Comment here is limited to noting some of the major groups of substances which have been considered in relation to the response of plants.

Organic acids. Numerous organic acids arise in anaerobic soils (Stevenson, 1967) and of these the volatile fatty acids are often the most abundant, especially acetic acid, but formic, propionic, butyric and valeric acids also occur; the quantity of these acids evolved per 100 g of waterlogged soils sometimes exceed 2×10^{-3} M when ample substrates are present. In addition aromatic acids can be present, for example, p-hydroxybenzoic, p-coumaric and vanillic acids (Wang *et al.*, 1967). Many other acids have been detected but by comparison with the volatile fatty and aromatic acids they appear to be of minor importance as phytotoxins. Despite the production of organic acids the pH of the soil does not change in consistent manner when anaerobic conditions develop. Many factors affect soil pH and may stabilize itclose to neutrality (E.W. Russell, 1973).

Hydrocarbon gases. The occurrence of methane in anaerobic soils has long been recognized and more recently ethylene and a number of the higher hydrocarbons have been identified (K.A. Smith and R.S. Russell, 1969; K.A. Smith and Restall, 1971). Concentrations of ethylene exceeding 10 ppm* have been found in waterlogged field soils (K.A. Smith and Dowdell, 1974).

The production of ethylene in anaerobic soils has attracted attention because it is also an endogenous growth regulator in plants and it can induce biological effects in very low concentrations. Moreover, the sensitivity with which it can be measured by gas chromatography, irrespective of the presence of other compounds, obviates analytical complications which often arise in the study of the organic products in comparable concentrations. Ethylene can be produced by numerous fungi (Bird and Lynch, 1974) and Mucor hiemalis can be largely responsible in at least some soils; methionine and glucose, which can arise from the decomposition of fresh organic material, being necessary substrates (Lynch, 1972). The process has a number of unexpected features. A restricted supply of oxygen in soil (less than c 0.01 bar) is normally necessary for ethylene to be produced but this requirement appears to relate to the release of substrates rather than to the production of ethylene from them; when ample substrates are added to well aerated

*Whereas some workers quote the composition of gas mixtures as concentrations (*e.g.*, per cent, or parts per million, as may be oonvenient) others quote partial pressures in bar (or atmosphere). 0.1 bar corresponding to ten per cent by volume in the gas phase. Measurements are here quoted in the terminology used by the authors.

soil abundant ethylene may be released (Lynch and Harper, 1974a). Moreover, the gas can be evolved not only from fungal hyphae but also from a diffusate to which they can give rise (Lynch, 1974). The quantities of ethylene produced can be inversely related to the growth rate of the fungus (Lynch and Harper, 1974b) and production may not cease if the temperature is raised sufficiently to kill the organisms (Lynch, 1975). This latter characteristic has encouraged the suggestion that ethylene is produced by anaerobic bacteria which survive higher temperatures (A.M. Smith and Cook, 1974); however, they appear to have little capability to do so.

Carbon dioxide. When gas exchange is restricted and anaerobic metabolism proceeds in the soil the concentration of carbon dioxide increases. However, its greater water solubility than oxygen, by a factor of about thirty, causes it to diffuse more readily in solution. The steep gradients of oxygen which can occur from the surface to the centre of water-filled zones are consequently not accompanied by corresponding steep gradients of carbon dioxide in the opposite direction (Greenwood, 1970). Concentrations in excess of five per cent by volume and exceptionally of over twice this magnitude have been reported in the zones of soil which roots explore (E.W. Russell, 1973). High concentrations of carbon dioxide can be toxic to plants, the effects often being generally similar to those caused by deficient oxygen. However, evidence reviewed by Grable (1966) and Kramer (1969) suggests that in anaerobic soils carbon dioxide is a minor source of injury to plants by comparison with the deficiency of oxygen. This conclusion received further support from the work of Williamson (1970); 20 per cent carbon dioxide in the presence of one per cent oxygen affected the growth of tobacco plants in a similar manner to one per cent oxygen in the absence of carbon dioxide; the injury caused by this latter treatment is discussed.

Sulphides. Hydrogen sulphide, which can be produced when the redox potential is very low (see Table 4.2) is very toxic to plants. However, if ferrous iron is present a highly insoluble sulphide is formed; sulphides are therefore unlikely to be toxic when soil contains appreciable quantities of soluble iron, but hydrogen sulphide has been considered to be a cause of injury in some soils (Vamos, 1964).

Losses of Soluble Compounds of Nitrogen

Under anaerobic conditions considerable quantities of nitrate can be lost from the soil both by denitrification and by leaching in drainage water. The latter prosess is independent of the partial pressure of oxygen in the soil but since anaerobic soils are in practice often waterlogged it is relevant to consider the two processes conjointly.

Leaching. Although nitrate is highly soluble and undergoes no significant interactions with the mineral phase of the soil, the extent to which it is removed by leaching can vary appreciably depending on soil structure. If water penetrates freely through large pores of cracks nitrate dissolved in water in the fine pores in the intervening solid phase may be lost relatively slowly; this can be important in conserving nitrate in some agricultural soils (Cunningham and Cooke, 1958). Thus, the amounts lost by leaching vary widely depending not only on the nitrate content of the soil and rainfall but on soil texture. Losses ranging from 5 kg to nearly 50 kg hard per year have been estimated depending on soil type in typical arable land in England (see Cooke 1976 for review).

Denitrification. The term denitrification describes the sequence of reduction processes which can convert nitrate to nitrate or nitrogen gas in anaerobic soil conditions, *viz.* NO_3, $NO_2 \rightarrow N_2O \rightarrow N_2$. In some usages the term is restricted to biological reduction mediated mainly by facultative anaerobes. In addition nitrate may react chemically with soil organic matter leading to the production of nitrogen gas (see Allison, 1973; Skinner, 1975 for review), but there is little evidence that this process is important in natural soils.

A large number of bacteria, mainly facultative anaerobes,

are capable of denitrification. Jordan *et al.*, (1967) found twenty-two isolates from a single soil which could reduce nitrate to nitrite but some had little, if any, ability to mediate the subsequent reduction of nitrite. When the supply of substrates is favourable, denitrification can occur very rapidly; thus in laboratory conditions the virtually complete denitrification of 300 ppm nitrate has been observed at 30°C; an approximate two-fold reduction in the rate of the process occurred with a 10°C drop in temperature (Cooper and R.L. Smith, 1963).

Before modern analytical methods became available the occurrence of denitrification could be inferred only from the indirect evidence that the total quantity of nitrogen lost from soil could not be accounted for by leaching or absorption by plants. Arnold (1954) showed by direct measurement that nitrous oxide can be generated in arable field soils; there is now good evidence that this happens commonly if substrates are abundant and anaerobic microsites occur (*e.g.*, Burford and Millington, 1968; Dowdell and K.A. Smith, 1974). Figure 4.2 from the last named workers shows results from a clay loam soil in southern England in which the water content was high throughout the period of observation. The extent of denitrification cannot at present be inferred from such measurements because there is insufficient information on the rates at which nitrous oxide diffused away from the sites of production. However, it is evident that denitrification took place down to an appreciable depth in the soil and that the process can increase when the temperature rises in spring. It is of interest that widely varying ratios of oxygen to nitrous oxide can be found in samples of soil gases or water taken only a short distance apart (c. 0.5 m) in the same field (Fig. 4.3); this is attributed to the production of nitrous oxide in localized anaerobic zones between which there may be appreciable air-filled spaces (see Fig. 4.1). Attention was first directed to this important question by Greenland (1962); using much simpler methods he deduced that denitrification could occur simultaneously with nitrification in wet soils which contained both anaerobic and aerobic zones.

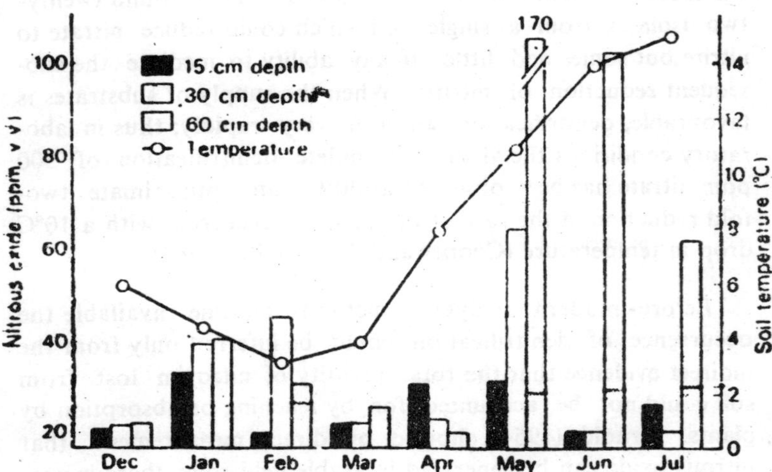

Fig. 4.2: Mean concentrations of nitrous oxide in clay loam soil (Evesham series) at three depths at different times of year and the mean soil temperature at 30 cm depth

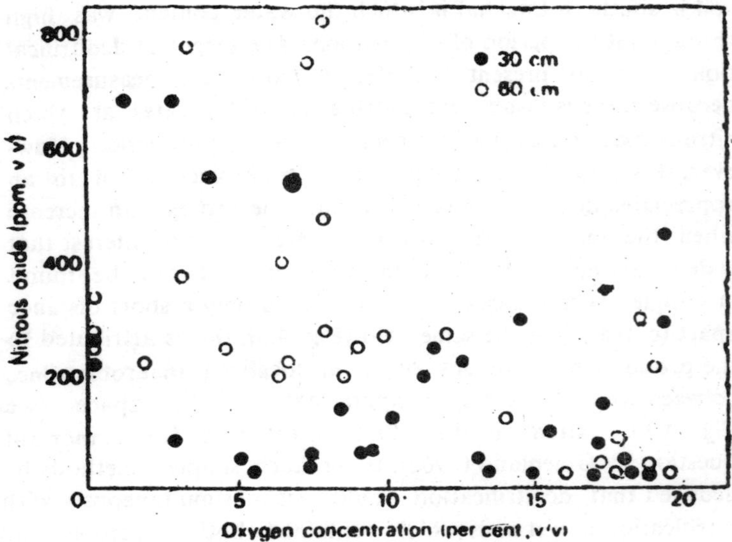

Fig. 4.3: Relationship between concentrations of oxygen and nitrous oxide at two depths in a clay loam soil (Evesham Series)

TABLE 4.3

Some Estimates of Losses of Fertilizer Nitrogen from Soil by Denitrification

Method	*Gasaous loss** (*per cent*)	*References*
Greenhouse pot experiments—^{15}N enriched fertilizer nitrogen	1-40	Broadbent and Clark (1965) (review)
Lysimeter experiments—loss estimated	4-42	Allison (1966)
Enclosed chambers—direct measurement of loss to atmosphere	<1-15 per week	Stefanson (1972) Martin and Ross (1968) Craswell and Martin (1975a)
Field experiments—loss of ^{15}N enriched nitrogen	2-36	Carter, Bennett and Pearson (1967) Myers and Paul (1971) Westerman, Kurtz and Hauck (1972) Craswell and Martin (1975b)

*Unrecovered N used as the sole estimate of loss of gaseous N except in enclosed growth chambers where loss was measured by the analysis of the evolved gases.

Some estimates of the extent to which nitrogen is lost from soil by denitrification are summarized in Table 4.3; the majority rest largely on indirect methods. The measurement of nitrogen gas as opposed to nitrous oxide which is produced by denitrification presents particular problems because of its presence in much larger concentrations in the atmosphere; the use of fertilizer labelled with ^{15}N reduces but does not entirely eliminate this problem as the atmosphere contains ^{15}N in low abundance. Experiments in enclosed chambers provide the most accurate estimates but the conditions under which they are conducted are difficult to relate to field conditions. Despite these uncertainties it appears that, depending on conditions, losses due to denitrification may range from negligible quantities to c. 20 to 45 per cent of the nitrogen applied to the soil. Denitrification may thus be a significant cause of losses of nitrogen in anaerobic conditions.

RESPONSE OF PLANTS TO ANAEROBIC SOILS

Species vary widely in their sensitivity to injury in anaerobic soils and in any one species the effect may change depending on its stage of growth and on the environment.

Morphological Effects

In sensitive plants, symptoms of injury can become evident rapidly, not only in roots but also in the shoots. Within a few hours leaves may wilt, or show epinasty (the downward curvature of petioles) (Fig. 4.4). Subsequent leaves can become chlorotic and senesce prematurely. The elongation of stems is often reduced and root growth is also much restricted. Exposure to anaerobic conditions for quite short periods, *e.g.*, twenty-four hours, can sometimes lead to permanent reduction in plant growth (Erickson and van Doren, 1961); longer exposure may lead to death. If plants survive anaerobiosis other morphological changes can occur, for example, the development of aerenchyma in roots.

Physiological Effects

The wide range of morphological changes in both roots and shoots caused by anaerobic conditions is indicative not only of the complexity of their physiological effects but also of the intimate co-ordination between the growth of roots and shoots. Some of the most common physiological effects are briefly discussed in the paragraphs which follow—many aspects are considered more fully by Grable (1966) and Kramer (1969) among others.

Permeability of roots. The seemingly paradoxical response that plants can wilt if the soil is saturated with water is attributable to the reduced permeability of roots due to the shortage of oxygen. In laboratory experiments with the root systems of tobacco (Nicotiana tabacum) a species sensitive to anaerobic soil conditions, Kramer and W.T. Jackson (1954) found that the permeability of roots to water could decrease by a factor of over three in a few hours. The ability of roots to absorb nutrients is also restricted but this usually does not have such an immediate effect.

Respiratory metabolism. Anaerobic conditions can cause a switch from aerobic to anaerobic metabolism with the consequent production of ethanol (Fulton and Erickson, 1964; Crawford, 1967). A comparison by the latter worker of the metabolism of plants which varied in their tolerance to waterlogging suggested that the concentrations of ethanol were higher in sensitive plants, this being associated with increased activity of alcohol dehydrogenase. Alternative respiratory pathways in resistant plants are discussed on page 104.

Endogenous hormones. As research has proceeded on the effects of anaerobic rooting media, increasing attention has been given to hormonal mechanisms. The production of both gibberellins and cytokinins in roots is now well established. Went (1938, 1943), who from many points of view initiated the modern study of growth regulators, suggested that inadequate aeration prevented the transfer from roots to shoots of growth factors' which were not then identified. More recently the reduction of the concentration of gibberellins in the xylem sap and in the roots and shoots of tomato plants has been demonstrated under anaerobic conditions (Reid *et al.*, 1969; Reid and Crozier, 1971). Moreover, the latter authors found that the inhibition of stem elongation caused by anaerobiosis in the root environment could be relieved by applying the synthetic gibberellin (GA_3) to shoots.

Burrows and Carr (1969) showed that the transport of cytokinins to shoots could be restricted under anaerobic conditions while Railton and Reid (1973) showed that the inhibition of stem growth and the senescence of leaves could sometimes be prevented by applying 6-benzyladenine (a synthetic cytokinin) to shoots; epinasty was also reduced. There is evidence also of the involvement of auxin in the response of plants to anaerobic soil conditions (Phillips, 1964 and b). Its concentration in shoots increases as epinasty develops. However, if the apex of the shoot is cut off, thus removing the major source of endogenous auxin, epinasty is reduced in plants grown in anaerobic soil; a causal relationship is suggested by the fact that the application of auxin to the cut shoot can reverse this effect. The production of abscisic acid can also be affected by anaerobic rooting media; Wright and Hiron (1972) found increased concentrations in shoots and considered this might be due to incipient wilting.

TABLE 4.4

Effects of Waterlogging for Four to Five Days on the Concentrations of Ethylene in the Shoot System of Various Species

	ppm in the gas phase (*v.v.*)			
	Tomato (*Lycopersicon esculentum Mill.*)	*Pea* (*Pisum sativum L.*)	*Field bean* (*Vicia faba L.*)	*Dwarf bean* (*Phaseolus vulgaris L.*)
Waterlogged	**1.17**	**0.93**	**0.50**	**1.36**
Control	**0.34**	**0.25**	**0.37**	**1.08**

However, ethylene is perhaps the endogenous growth substance which appears most closely associated with the response of plants to anaerobic soil; its concentration throughout susceptible plants often increases markedly (Kawase, 1972, 1976; El-Beltagy and Hall, 1974; M.B. Jackson and Campbell, 1974, 1976a). Some results of the last named workers are shown in Table 4.4. Moreover, exogenous ethylene can cause epinasty in shoots (Leather *et al.*, 1972) and, if the supply of oxygen in the rooting medium of tomato plants (Lycopersicon esculentum) is reduced, the concentration of ethylene in leaves and also the degree of epinastic curvature can increase comparably (M.B. Jackson and Campbell, 1976a). However, many aspects of the manner in which a restricted supply of oxygen affects plant growth control mechanisms remain obscure.

Relative Significance of Restricted Oxygen Supply and Toxic Substances

It is implicit in earlier discussion that anaerobic soils can be an unfavourable environment for plant growth both because of low oxygen tension and because of the accumulation of toxic substances. What is the relative significance of these two potential causes of injury? An attempt to answer this question illustrates the meagreness of present information on many aspects and also the variability of the situations which can occur.

The supply of oxygen. At the outset the difficulty should be recognized of estimating the supply of oxygen actually available to plant roots in soil, except in a very approximate manner. Among the reasons are:

1. As shown earlier in this chapter considerable gradients in oxygen can occur when anaerobic conditions develop in the soil (Fig. 4.1). On these grounds alone the partial pressure of oxygen at the surface of roots may be expected to differ from the mean for any appreciable volume of the soil.
2. The consumption of oxygen by rhizosphere organisms may lower the partial pressure at the surface of roots to an extent which cannot be directly measured.
3. The soil is not the sole source of oxygen to roots. There is abundant evidence of the longitudinal transport of oxygen in the air spaces from shoots to roots of typical mesophytic species (Evans and Ebert, 1960; Barber *et al.*, 1962; Jensen *et al.*, 1967; Greenwood, 1969). The quantities transported are not sufficient to supply the requirements of plants except in some species habituated in anaerobic soils, but it seems

> probable that longitudinal transport may much modify the supply of oxygen to different parts of root systems when the soil contains anaerobic zones. Greenwood (1969) found that appreciable quantities of oxygen were transported through roots 5 cm long and released in the outer medium near the root tip. This suggests that roots in oxygenfree zones of soil may receive oxygen from neighbouring aerated regions.

For the foregoing reasons estimates of the partial pressure of oxygen necessary in soil to sustain root growth rest largely on indirect evidence. On theoretical grounds, supported by experiments, Greenwood (1969) concluded that the affinity of cytochrome oxidase for oxygen is such that aerobic respiration is unlikely to be impeded unless the oxygen concentration at the root surface is approaching zero; a partial pressure of 0.01, bar, as opposed to 0.21 bar in the normal atmosphere, was considered to be adequate.

There is, however, much evidence of variation in the sensitivity of different species to low concentrations of oxygen. Most estimates of the minimum partial pressure of oxygen which does not appreciably restrict root growth, range from 0.2 to c. 10 bar (see Carr, 1961; Greenwood, 1969 for review). Part of the variation is undoubtedly due to differences in experimental conditions; at low temperatures, the partial pressure of oxygen which can sustain growth may be considerably reduced (Carr, 1961). Moreover, in some experiments the possibility cannot be excluded that root growth was also affected by toxic substances produced in the soil. Interest therefore attaches to the work of Williamson (1970) who studied the response of tobacco (Nicotiana tabacum), one of the crops most sensitive to anaerobiosis. Oxygen was excluded from the soil either by saturating it with deoxygenated water or by forcing oxygen-free nitrogen through it. Table 4.5 shows that the absence of oxygen induced in either way has very severe and similar effects. Since the passage of nitrogen through the soil should have reduced the concentration of any volatile toxic products there can be little doubt that they did not make a large contribution to the injury observed in soil saturated with water; the absence of oxygen was the evident cause. The marked reduction of growth which was observed when one per cent oxygen in nitrogen was passed through the soil must be attributed to the same cause.

TABLE 4.5

Effect of Gas Mixture Containing One or Zero Per cent Oxygen Which was Forced Through Soil, or Flooding with Deoxygenated Water, On the Growth of Tobacco Plants (Nicotiana Tabacum). These Treatments were Applied for 24 or 48 h and the Plants were Examined Five Days After the Commencement of the Experiment. Mean Results Expressed As Per cent of the Values for Control Plants Grown in Aerated Soil

Treatment	*Root dry weight*	*Shoot weight*	*Leaf area*
		per cent of control	
Gas mixture 1 per cent O_2: 99 per cent N_2	49	100	79
0 per cent O_2 100 per cent N_2	12	59	38
Flooded with deoxygenated water	15	54	39

Writing of United Kingdom conditions greenwood (1969) concluded that root growth is unlikely to be impeded through lack of oxygen unless the soil is almost completely waterlogged and contains some oxygen-free zones. Present information does not disprove this generalization.

Effects of toxic substances produced in soil. Quantitative estimates of the magnitude of injury caused by toxic metabolities in anaerobic soils are beset by difficulties broadly comparable to those which arise in assessing the effects of the reduced partial pressure of oxygen; the localized production of toxins in anaerobic zones can cause variations in their distribution and the possibility of their metabolism by rhizosphere organisms may cause the concentration experienced by roots to differ from that in their near vicinity.

Uncertainty regarding the significance of ethylene produced in soil illustrated this problem. The view that it might play a major role in the injury of plants in anaerobic soil was encouraged by several observations. Whereas concentrations equivalent to over 10 ppm can occur in anaerobic soils (Dowdell and K.A. Smith, 1974), 2 ppm or less can restrict root extension in some species (K.A. Smith and Robertson, 1971), and, the tolerance of a number of species to externally applied ethylene can vary in a manner broadly comparable to their tolerance of anaerobic soils. Beyond this, some of the symptoms of injury from anaerobiosis can result from increasing concentrations of ethylene within the plant.

None the less, exogenous concentrations of ethylene which seriously retard root extension in barley (Hordeum vulgare) do not induce the typical symptoms of injury by anaerobiosis. Even when root extension is arrested by treatment with the gas for ten days or more, and at concentrations much in excess of those found in the soil, shoot growth can be unaffected and if ethylene is removed from the root environment, root elongation can soon recommence (Crossett and Campbell, 1975). However, since ethylene which enters the roots of tomato can move upwards to shoots (M.B. Jackson and Cambell, 1975), exogenous production of the gas in anaerobic soil conditions seems likely to make some contribution, though possibly only a minor one to the concentration within plants and hence to the injury which they sustain.

The toxicity of volatile fatty acids has been extensively studied in rice soils (see Stevenson, 1967), E.W. Russell (1973) quotes estimates that 10 ^{2}M acetic acid and 10^{-4} M butyric acid are injurious to rice and that, after readily decomposable organic matter has been added to the soil, these concentrations may be exceeded to an extent which makes the land unsuitable for planting the crop until the acids have been largely removed by leaching or other processes. These observations are perhaps the strongest evidence of volative fatty acids being a major cause of injury to plants in anaerobic field soils. There is evidence also that these substances contribute to the creation of unfavourable conditions when massive quantities of animal waste are added to the soil (Copper and Comforth, 1975).

The effects of phenolic acids derived from plant debris, including stubble mulches, have been considered particularly by Tousson *et al.*, (1968) and McCalla and Norstadt (1974). The former workers concluded that these compounds were responsible for some 60 per cent of the toxicity in either extracts from such soils. McCalla and Norstadt have attributed toxicity both to phenoxy acids and to patulin, an antibiotic, but there appears to be little information on the concentrations in which this latter substance occurs.

There thus appears little doubt that when appreciable quantities of fresh organic debris are present in anaerobic soils, especially in the close vicinity of germinating seed products of their decomposition can contribute to the injury of plants. The paucity of definite information at the present time is perhaps best regarded more as a reflection of the lack of detailed information on this difficult subject than as evidence that toxic products are usually of negligible significance. Grable (1966) stated 'Few areas of scientific investigation are as unintelligible as the relationship between soil physical properties, aeration and plant growth. This is not surprising, however; few other areas are so experimentally complex and important variables so often confounded or ignored.' A decade's intensive work has not outdated this conclusion; indeed, as research has proceeded the complexity of the subject has become more evident.

Tolerance of Anaerobic Conditions

Both morphological and physiological characteristics of plants can confer tolerance to anaerobic conditions. One of the most obvious anatomical adaptations is a considerable increase in the intercellular air spaces in the cortex, thus providing canals parallel to the axis of the root through which gases can diffuse longitudinally. This aerenchyma is particularly well developed in rice (Oryza species) and numerous aquatic plants (Arikado, 1955; Laing 1940; Coult, 1964). In such species it is able to supply much, if not all, of the oxygen requirement of roots and indeed also provides it to the surrounding rhizosphere soil (Bidwell *et al.*, 1968). Another morphological feature which may favour survival is the ability of plants rapidly to develop adventitious roots close to the soil surface where the oxygen tension is usually higher or more quickly restored after transient waterlogging; this appears, for example, to be important with tomatoes (W.T. Jackson, 1955).

One of the physiological characteristics which has received the greatest attention is the development of alternative respiratory pathways, thus restricting the production of ethanol. Organic acids, for example malic and shikimic, may be instead produced (Crawford and Tyler, 1969; Tyler and Crawford, 1970). Numerous indications of the involvement of growth regulatory substances in the response of plants to anaerobic soils encourage the speculation that differences in the endogenous hormone patterns may influence tolerance—but this is at present an unexplored field.

SOME AGRICULTURAL ASPECTS

The complexity of the factors which determine the rapidity with which anaerobic conditions develop and toxins are produced, when soil has become saturated with water, are evident from the first part of this chapter. Subsequently, the equally diverse ways in which plants respond were illustrated. Even if these processes were much more fully understood, widely applicable generalization would be difficult. Discussion is here limited to two aspects—namely commonly observed effects of transient waterlogging and factors which may influence the subsequent recovery of plants.

Transient Waterlogging of Crops

Except in limited circumstances, of which the cultivation of paddy rice is the obvious example, agricultural problems due to anaerobic soil conitions are usually transient, being the consequence of the soil being nearly or completely waterlogged for periods ranging from days to weeks, when precipitation is high and evaportranspiration low; at other times of the year water may be deficient.

The injury caused by transient waterlogging can very much depending on the stage of growth when it is experienced. Part of this variation may be due to seasonal differences in oxygen consumption (see Table 4.1) in the soil which influence the rate at which anaerobic conditions develop, but there is much evidence of changes in the sensitivity of plants during their life cycle. Germinating seeds are commonly regarded as particularly sensitive to injury. They are totally dependent on the surrounding soil for oxygen; the transfer of oxygen from the atmosphere through the plant can occur only after shoots emerge. Damage to early seeding growth in waterlogged soil can be exacerbated if they are in close proximity to plant debris of recent origin which is saturated with water (Tousson *et al.*, 1968; McCalla and Norstadt, 1974).

It is commonly observed that after plant shoots have emerged from the soil their tolerance to transient waterlogging can very greatly. Guyon (1970) quotes experiments which indicate that waterlogging for three days was without effect on winter cereals until February (in the northern hemisphere) but in May or June it led to a 20 per cent decrease in yield, while waterlogging for eighteen days in winter caused about a 20 per cent loss of yield and in May or June it resulted in premature death. Qualitatively similar, though often less extreme, changes in the sensitivity of cereals to waterlogging have been found in other studies (Ikeda *et al.*, 1957; Aleksandrova and Skazkin, 1964; Swartz, 1966). In general, it appears that the yield of cereals is depressed to the greatest extent if the soil is waterlogged when the reproductive organs are developing. Erickson and van Doren (1961) reached similar conclusions with paes; waterlogging for twenty-four hours causing a considerable reduction in yield. Changes in sensitivity during the growth of this crop are illustrated. However, in maize Lal and Taylor (1969) found transient waterlogging to be particularly damaging early in the season.

The fact that very short experiences of waterlogging for example twenty-four hours, can sometimes cause permanent injury to plants suggests that transient anaerobiosis may sometimes have a significant effect when a conventional visual examination of the soil provides no indication that reducing conditions have occurred. Since the oxygen in the soil may be able to sustain respiration at the normal rate for only one or two days it seems possible that the sealing of the soil surface, as the result, for example, of slaking in heavy rain, map cause at least localized anaerobic zones to develop in normally well drained soils, during the course of even a day, if the rate of respiration is high. In tropical situations visual symptoms comparable to those caused by anaerobiosis are sometimes seen in plants in these circumstances but no definite conclusion is possible in the absence of information on the composition of the soil atmosphere. These situations appear to be particularly deserving of further study.

Recovery of Waterlogged Crops

If plants survive partially anaerobic conditions for an appreciable period two factors may exert a particular influence on their subsequent growth. Living roots are likely to be confined to the surface layers of the soil in which the oxygen supply is usually greater and the supply of nutrients in the soil may be modified.

On the former question there are few quantative field observations though the restriction of roof growth when the partial pressure of oxygen falls appreciably below c. 0.1 bar is well documented, and in the field few, if any, roots are usually observed more than a few centimetres below a static watertable.

Thus, if the watertable subsequently falls rapidly, as is possible in a soil of high hydraulic conductivity, the rate of root extension may be a major factor which influences the ability of the plant to survive.

Crops growing in soil which has been waterlogged often have the pale green colour commonly attributed to an inadequate supply of nitrogen and may respond rapidly to the application of that nutrient; the losses of nitrogen from soil which can be caused by leaching and denitrification provide a ready explanation. The view that nitrate may be more beneficial than ammonium under these circumstances is encouraged by the conclusions from some loboratory studies that if oxygen is in limiting supply, the combined oxygen in the nitrate ion may act as an electron acceptor within plants (Arnon, 1937, 1939; Siegel, 1961). However, at present there is no definite evidence that this effect is usually important in the soil. Information reviewed by Grable (1966) suggests that the relative benefit of nitrogen in the form of nitrate or ammonium may depend on several factors, including the nitrogen requirement of the plant and the supply of trace elements.

Finally, with regard to recovery from waterlogging injury the apparently beneficial effects which have been reported from the application of some growth regulators to shoots should be recalled but whether any such treatment can in practice offset the injury of sensitive crops remains a subject for speculation.

5

Utilization of Water

In many parts of the world an inadequate supply of water is the most conspicuous and continuing limitation to crop growth. Water relations thus occupy a central place in any survey of developmental physiology. But much of this subject lies outside the scope of a discussion of root function in the soil.

Meteorological factors and the characteritics of plant shoots can greatly influence the effects on crops of variations in water supply. These are referred to only briefly in this chapter and attention is directed mainly to the absorption of water by root systems, to the effects of variations in water supply on their growth and function and to their morphological features which contribute to the survival of plants when the supply of water is limiting. The discussion of a number of aspects which require the joint considera-tion of characteristics of the soil and of the plant is deferred to Part II of this book, for example the effects of water-logging and the transfer of water across the root/soil interface.

THE DRIVING FORCE FOR THE MOVEMENT OF WATER

If the water economy of plants is to be seen in perspective, account must be taken of their evolutionary history. They originated in an aquatic environment and photosynthesis, the unique metabolic process on which plant growth depends, requires the absorption of carbon dioxide from the environment with the simultaneous release of oxygen. In submerged plants this exchange occurs in solution but when terrestrial plants evolved, the site of photosynthesis being necessarily in the above-ground parts, the exchange of carbon dioxide and oxygen with an unsaturated atmosphere had the inescapable consequence that water vapour was simultaneously lost. No membrane system was evolved—nor has synthetic chemistry yet produced one which permits the free exchanges of oxygen and carbon dioxide between an aqueous phase in plant cells and an unsaturated atmosphere without the loss of water. Transpiration is thus an inescapable physical consequence of the development of terrestrial vegetation.

Studies of the utilization of water by crops entered a new phase with the work of Penman (1948). Earlier discussions had frequently been concerned with the transpiration coefficient; that is to say the quantity of water used per unit of dry matter produced, a physiological relationship between growth and water loss often being implied. Penman, however, showed that, during summer in temperatre regions, the quantity of water lost from a complete ground cover of short grass, which was well supplied with water, is about 0.8 of the water loss from the same area of a free water surface; in spring or autumn the corresponding figure was somewhat lower, *c.* 0.7 (Penman, 1948). Somewhat different values apply to other types of ground cover and Penman's work provided a valuable and widely used basis for estimating the water requirements of growing crops (see Penman 1963 for fuller discussion).

Transpiration by leaves in thus the driving force for the movement of water through plants. The simplicity of this relationship should not however, obscure the complexity of processes which can regulate the movement of water within plants, especially when they are subject to a degree of water stress. This matter can be meaningfully discussed only with a thermodynamic background which is here considered only briefly; Slatyer (1967): Kramer (1969); Kozlowski (1968, 1972) and Weatherley (1970) are among the standard sources from which fuller information can be obtained.

WATER POTENTIAL

The movement of water depends on the existence of gradients of decreasing water potential—that is to say its free energy—which bears no direct relationship to the total quantity of water present. The transfer of water from the soil through plants to the atmosphere is thus appropriately considered in terms of potential gradients. None the less the quantity of water present exerts an important indirect effect; it largely determines how the potential changes when water is introduced into the system or withdrawn.

The same basic thermodynamic principles apply to the movement of nutrient ions into plants. However, in the majority of circumstances it is adequate to consider only their concentration in the solution with which roots are in contact, since it is usually sufficiently dilute for the chemical activity of an ion to vary closely with its concentration.

PASSAGE OF WATER THROUGH THE PLANT

Resistances to Flow

The smaller loss of water from a complete vegetation cover than from a free water surface, as shown by Penman reflects resistances to the flow of water along its pathway from the soil to the transpring leaves. Ohm's law provides a convenient analogy which has been widely used since van den Honert (1948) drew attention to its usefulness; if water moves at a constant rate through a system, the ratio of the drop in water potential to the resistance to water flow at each step in the transfer process will be constant. Thus, the relative magnitudes of the resistances to water movement along the soil-plant-atmosphere pathway, under constant conditions, can be derived from the change in potential. To describe the passage of water realistically the simple Ohm's law analogy must, however, be amplified in two respects. *First*, the major resistances to water flow in the plant can vary in a complex manner. *Second* account must be taken of the capacity of the system, to use another electrical analogy; if increased transpiration leads to a lowering of the water potential in any part of the transpiration stream, water will be withdrawn from the surrounding cells with a corresponding drop in their water potential.

It does not, however, indicate all processes which cause water to move in plants. Although transpiration is dominant under normal circumstances it is not the sole cause of water movement. If plants are maintained in a humid atmosphere the flow of water through them does not usually cease completely; in some species droplets of water may be released by guttation through hydathodes at leaf tips. This process is attributed to gradients in water potential created by osmotic forces; the bleeding sap which can pass out of cut stems or roots, which is sometimes regarded as evidence of the 'active' uptake of water is similarly explained. These processes are not considered further as they account for only a very small fraction of the total water which moves through transpiring plants.

Movement of Water Through Plant Shoots to the Atmosphere

When plants are transpiring in a relatively dry atmosphere the decrease in water potential between the interior of the leaf and air exceeds by a considerable factor that in any other part of the system. Between the soil and the transpiring leaf the decrease in water potential is unlikely to exceed about 50 bar in an unwilted plant and is often very considerably less, but the drop is water potential from the interior of the leaf to the atmosphere may approach 1000 bar. It follows, therefore, that the highest resistance to water movement in a plant system can occur in the leaf.

The movement of water across the cuticle on the leaf surface is small except in plants habituated to constantly humid environments and the major gas exchange occurs through the stomata. The finely balanced mechanisms which control the movement of the guard cells have been the subject of considerable research (see, for example, Meidner and Mansfield, 1968; Monteith and Weatherley, 1976). The two main circumstances which usually lead to stomatal closure are an increase in the concentration of carbon dioxide in the gaseous phase within leaves and a decrease in water potential of leaf cells; these two responses are believed to be linked by a hormonal process in which abscisic acid is involved. The role of stomata in water conservation is thus evident; when decreasing leaf water potential causes stomatal closure, the loss of water can be much less than that which the prevailing meteorological factors would permit (Lake *et al.*, 1969). However, the efficiency of the control mechanism varies widely between species and it reduces water loss to a negligible level only in xerophytes. The spraying of plants with artificial 'antitranspirants' can restrict water loss, the action of some such substances being to modify stomatal action.

The resistance to the upward movement of water through the xylem of the stem and leaf traces is small beside that across the leaf surface. This is well illustrated by the fact that the rate rate of transpiration is little affected when an appreciable part of the xylem in the stem has been severed by transverse cuts (Mackay and Weatherley, 1973). It is often assumed that the resistance to the passage of water across the leaf varies only because of stomatal movement but there is some evidence that the resistance in the mesophyll may also decrease if the rate of transpiration rises (Boyer, 1974).

Transfer of Water Across Roots

The main resistance in roots to the movement of water is in its radial passage across the cortex to the xylem. Next to the resistance in leaves it is the principal barrier to water movement throughout the plant; by comparison resistance in the xylem of the root is usually small. This can be easily demonstrated by cutting off roots of plants which are immersed in a solution;

transpiration may then increase because water can move directly into the cut ends of the conducting tissue. None the less the resistance in the xylem of the root is not necessarily negligible. Passioura (1972) has suggested that, when grasses are absorbing water mainly from deep in the sub-soil, viscous resistance of water flow in the conducting tissues of their thin roots can be important. Newman (1974) cites evidence which points to the same conclusion of roots exceed *c.* 1 m in length.

It is usually considered that water crosses the cortex of roots to the endodermis mainly, if not entirely, in the free space (Weatherley, 1963; Tanton and Crowdy, 1972), but some workers consider that cytoplasmic movement, or transfer between vacuoles in cortical cells, may make some contribution (Newman, 1974).

When plants of different species are grown in the same environment considerable contrasts in the permeability of roots to water have been detected (Newman, 1973) and there is much evidence that the resistance to the radial transfer of water across a root can vary with environmental factors. Decreased transpiration caused by anaerobiosis, or unfavourable temperature in the rooting medium, has been attributed to the reduction in the permeability of membranes; the application of respiratory inhibitors can have a similar effect. A review of the effects of temperature by Kramer (1969) indicates that the restriction of absorption caused by low root temperatures is greatest in plants adapted to warm habitats.

Variations in the rate of transpiration may also modify the resistance to the radial transfer of water across roots. Weatherley and his co-workers (Tinklin and Weatherley, 1968: Stoker and Weatherley, 1971) found in several species that if transpiration is increased by a reduction in the humidity of the atmosphere, the resistance to the movement of water across the root decreased. This effect has not, however, been found by all investigators (*e.g.*, Neumann *et al.*, 1974) and some observations suggest that it may occur only at a limited range of transpiration rates (Hailey *et al.*, 1973). Differences in experimental methods may perhaps help to explain discrepancies between investigations.

Comparative studies of the entry of water into different parts of root systems provide additional evidence that the radia movement of water across roots is influenced by the rate of transpiration. When the supply of water is ample it enters more rapidly into the apical zones of roots where the endodermis is not suberized (Hayward *et al.*, 1942; Graham *et al.*, 1974). However, experiments by Brouwer (1953) indicate that changes in the rate at which water is absorbed can modify the extent to which it enters different parts of young roots. The uptake of water by a single root of broad bean (*Vicia faba*) was enhanced either by transferring the plant from dark to light or by placing the other roots in an osmolyte of osmotic potential *c.*—2.0 bar. Both treatments led to considerably more rapid uptake of water, especially 7.6—12.5 cm from the apex; absorption by them then exceeded that in the special 2.5 cm in which the rate of uptake had been greater when transpiration was low. Newman (1974) has reviewed other evidence that the entry of water into different parts of young roots can be varied by external factors but this subject is still very incompletely understood even when the entire root system experiences the same water potential. When plants receive a restricted water supply in soil, further complications may be introduced as a result of discontinuous contact between roots and the soil.

None the less, all investigations of the uptake of water by different parts of the root system indicate that water can enter most readily into the young unsuberized zones of roots, but absorption is not confined to them. In woody perennial plants, of which over 99 per cent of the root system may be suberized, slow abrorption through the order tissues is likely to make a major contribution to the total water up-take (Kramer and Bullock, 1966)

Sources or more detailed information on the water relations

of plants include Kramer (1969); Hsiao (1973); Newman (1974) and Monteith and Weatherley (1976).

EFFECTS OF THE EXTERNAL SUPPLY OF WATER ON THE FUNCTION AND GROWTH OF ROOT SYSTEMS

In the preceding chapter the effects of the supply of nutrients on absorption by roots and on the form of root systems was discussed primarily on the basis of observations when roots were grown in solution culture and no other factors were varied. There is no corresponding simple and direct way for studying the effects of differenes in the external water potential; it cannot be varied in controlled artifical systems with the same facility as the concentration of nutrients. Evidence on the effects ow water potential on root function comes mainly from experiments in soil. Thus there is some overlap between this chapter and those in Part II of this book, which are more directly concerned with soil conditions.

General Relationships between Water Potential and Root Function

The terms permanent wilting point 'and' wilting percentege are used to decribe the maximum water potential, or water content, in soil which causes plants to wilt. Slatyer (1967) has pointed out that the permanent wilting point should not be regarded as a soil constant since wilting depends on the loss of turgor from leaves and it is influenced by the rate of transpiration as well as by the osmotic pressure of the vacuolar sap of cells. Beyond this the practical usefulness of the wilting point concept in field situations is often limited because ample water in part of the rooting zone may permit growth to continue although the soil adjacent to much of the root system is close to the wilting point.

It is generally considered that a water potential of 15 bar throughout the rooting zone would lead to the permanent wilting of the majority of crop plants (Kramer, 1969) but the rate of root extension starts to decrease at much higher water potential *e.g.*, often about *c*—0.5 bar though root extension may continue slowly until the water potential falls to—10 bar or lower (Lawlor, 1973; Newman, 1966). Some experiments with maize (corn, Zea mays) and tomatoes (Lycopersicon esculentum) indeed indicate that if part of a root system is well supplied with water, some growth can still occur in roots which are surrounded by soil in which the water potential is less than —40 bar; they may remain alive at appreciably lower water potentials (Portas and Taylor, 1976).

When osmotic forces contribute appreciably to the water potential, for example in saline soils, or after large addition of fertilizer, the restriction of growth may be greater than if low soil water potential arises from matric forces alone Osmotic potentials *c*—2 to —4 bar can be injurious to many plants (Kramer, 1969), but halophyles are more tolerant. The injury caused by low osmotic potential may be associated with interference with hormonal processes.

Because of the experimental problems of varying the water potential uniformly through the rooting zone, other factors being maintained constant, the effects of water potential on nutrient uptake have been little studied independently of those on growth. However, if low water potential reduces root growth a restriction of nutrient uptake is to be expected because of the reduced metabolic activity of the roots, moreover the metabolic demand of the whole plant for nutrients can be reduced. If osmotic as opposed to matric forces are responsible for low water potential, nutrient uptake may also be affected by competitive processes for example, in saline soils a high concentration of sodium can depress the absorption of other cations.

Effects of Gradients in Water Potential Within the Rooting Zone

Except when the soil is close to or above field capacity appreciable gradients in water potential usually occur. When water is lost in dry weather its potential usually decreases more rapidly near the surface than in the deeper soil layers.

This is due not only to the downward movement of water, but also to its evaporation into the atmosphere and absorption by roots which are often more abundant near the surface. The gradient can be enhanced if gravitational drainage is is slower in the deeper layers and sometimes also because their capacity for water retention is higher.

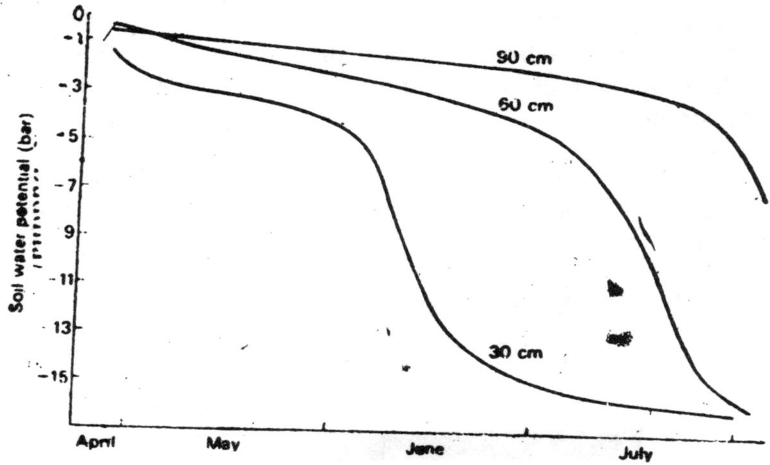

Fig. 5.1: Changes in water potential at three depths in a light alluvial soil underlying an established pasture during three summer months in which the pasture was sheltered from rain. The water potential at 15 cm different little from that at 30 cm during the major part of the period

Climate, the distribution of roots and soil characteristics all interact in determining the magnitude of gradients of water potential in the soil under growing plants. Thus, the relationship observed in any set of circumstances cannot be regarded as widely representative. None the less Fig. 5.1, constructed from the work of Garwood and Williams (1967a),* illustrates the types of situation which can develop in dry summer weather in a temperate locality; during three summer

* The author is indebted to Drs Garwood and Williams for making available unpublished information which allows soil water potentials to be derived from the measurements of gravimetric water content given in the reference cited.

months a sward of perennial ryegrass (Lolium perenne) in southern England was protected from rain by a transparent cover at sufficient height to allow free air circulation. A period of drought was thus reasonably simulated. The water potential fell considerably more rapidly at and above 30 cm than at greater depth, and as time passed the uptake of water became increasingly dependent of roots which had penetrated deeply. The general manner in which these conditions affect the distribution of roots and the absorption of nutrients will now be considered.

Root distribution. Under field conditions variations in water supply are frequently the major cause of differences in the distributed of roots, particularly the depth they attain in soil. It illustrates contrasing rooting patterns in comparable soils due to climatic differences while Fig. 5.2. shows variations.

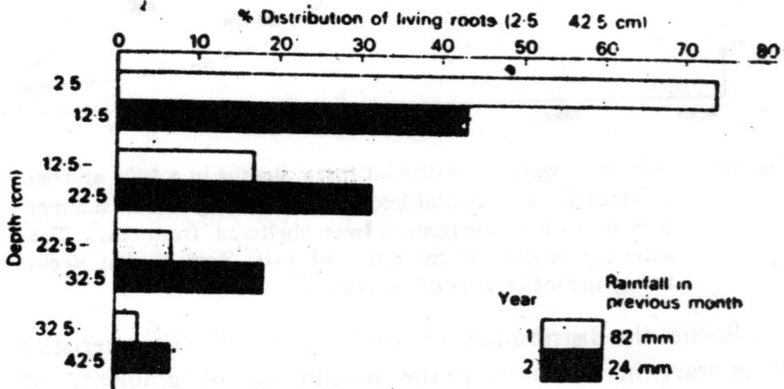

Fig. 5.2: Constrasting distribution of living roots of spring burley grown in the same field in successive years: measurements two months after planting. In Year 2 driver conditions in the surfce soil much reduced the fraction of roots in that zone.

The relationship between root density and depth was exponential in Year 1 and linear in Year 2. The percentage of the total variation accounted for on the two bases was : Exponential relationship year 1 . 98 (13), year 2 : 74). (25). Linear relationship—year 1 : 78 (55), year 2 : 99(7). The figures in brackets below the maximum percentage difference between the calculated values and those absorved for individual horizons caused by differences in rainfall on the same field in two successive years. This variability is another example of the capacity of roots of 'compensatory growth' a restriction of growth in part of the root system may lead to increased growth of more favourably placed root members.

The analyse adequately the effects of water supply on root growth in the soil, account must be taken not only of the

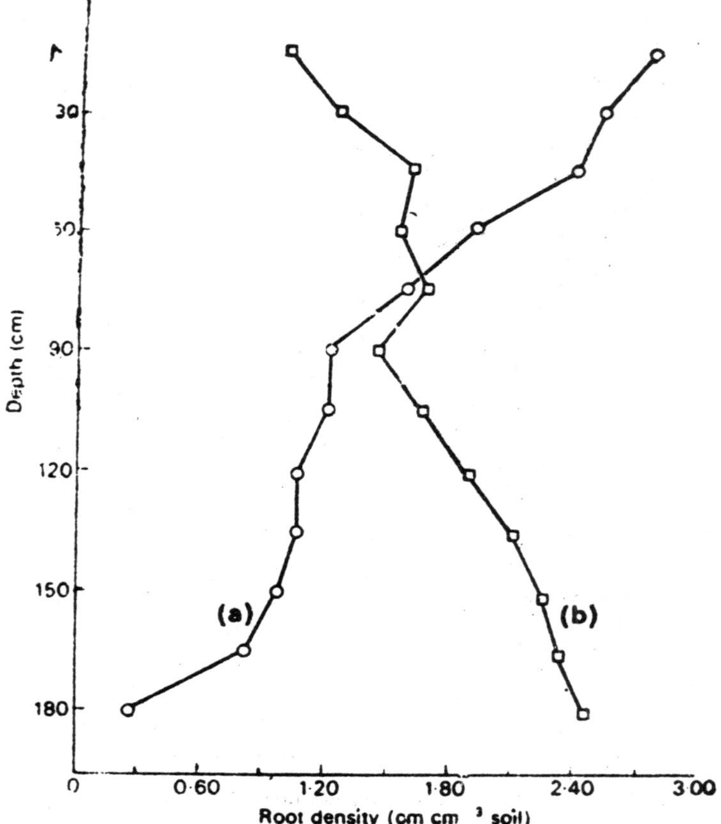

Fig. 5.3: Eeffect of decreasing water potential for three weeks specially in the appear soil on root distribution in cotton plants (Gossypium hirsutum). (a) at beginning of experiment: (b) at end, changes in soil water content and protential are shown in Fig, 5.4

water potential at different depths but also of the rate at which water moves through the roil towards roots; the determines the extent to which water potential is maintained adjacent to the root as absorption proceeds. The difficulty of measuring the potential and the quantity of water in soils simultaneously with root distribution has restricted detailed studies of this subject. The Rhizotron at Auburn. Alabama, in which the growth of roots observed in large volumes of soil against glass windows has, however, provided interesting information Figure 5.3 shows the distribution of the roots of cotton (Gossypium hirsutum) grown in a uniform fine, loamy send of relatively low water retention at the beginning and the end of a twenty-one day period in which no water was added to the soil (Klepper *et al.*, 1973). Initially when the water potential throughout the rooting depth was c.—0.1 bar the density of roots decreased from the surface downwards. However, during the during cycle when water was lost considerably more rapidly from the surface layers (Fig. 5.4 and b), the pattern of root distribution was reversed (Fig. 5.4). The density of roots in the dry surface soil decreased due to death by desiccation, while a considerable proliferation occurred in the deeper layers in which the soil water potential fell little until late in the experiment. When plants were grown simultaneously in indentical conditions, except that an ample water supply was maintained throughout the rooting zone, there was little change in distribution of roots during the three week periods. The effects of the desiccation of the upper part of the root system shown in Fig 5.3 are basically similar to those which can be induced by lowering the solution level when plants are grown in solution culture The continuing—often enhanced—growth of the lower parts of a root system when water shortage inhibits growth nearer the soil surface has also been demonstrated by Newman (1966); however a response as great as that shown in Fig. 5.3 seldom occurs under field conditions.

Nutrient uptake. Quite short periods of warm, dry weather can cause large variations in the absorption of nutrients from near the soil surface. This is conveniently

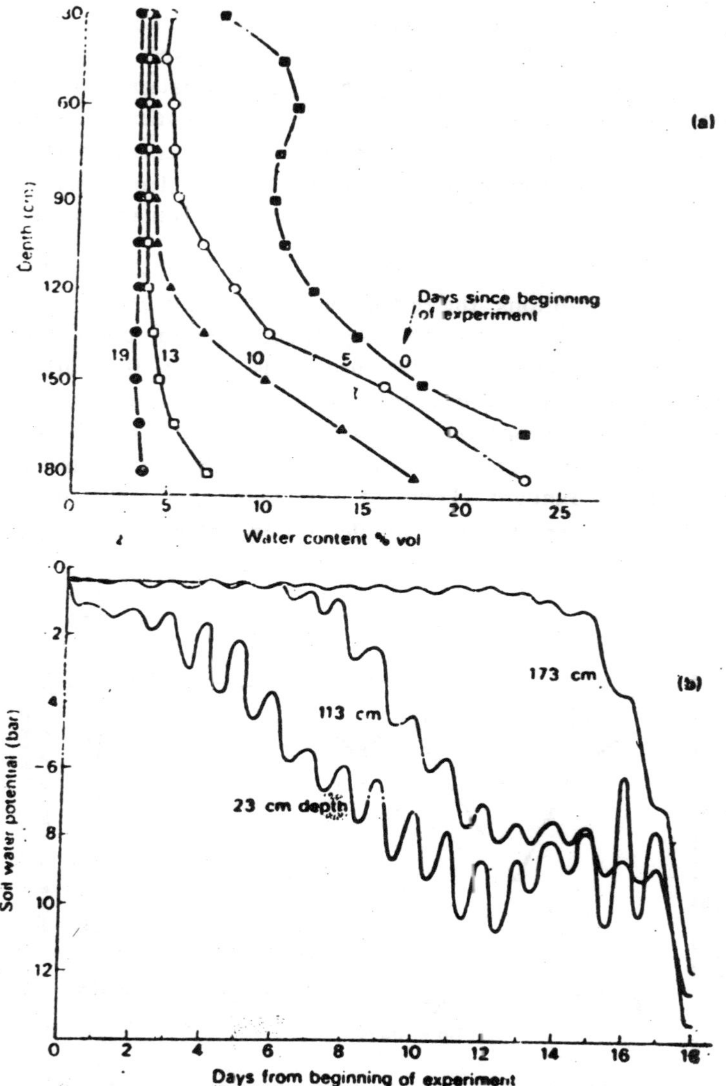

Fig. 5.4: Changes in soil water content (a) and soil water potential (b) at different depths in the soil during the experiment in which root distribution was illustrated in Fig. 5.3. Diurnal variations in soil water potential maximum at 06.00 hours) are evident

illustrated by experiments in which soil at different depths has been labelled with radioactive tracers; they enable uptake from different depths to be compared. Figure 5.5 gives the results of an experiment conducted in this way; variations in the moisture content at the depth of 5 cm in soil under a sward of perennial ryegrass (Lolium perenne) led to corresponding variations in the absorption of calcium from that depth relative to the deeper soil layers, phosphate showed a similar response (Newbould, 1969).

In dry conditions the restriction of nutrient uptake from near the soil surface can impose a particular restraint on plant growth because, when the water supplys is ample, the surface layers of soil are not infrequently the major source of nutrients for many annual crops as well as perennial grasses; not only is the density of roots greatest in this zone but also nutrients are usually most aundant. Thus, if the water potential of the surface soil decreases, but ample water remains accessible

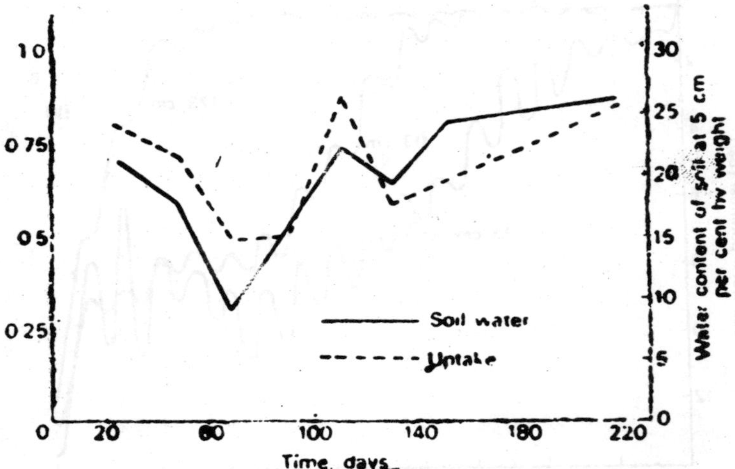

Fig. 5.5: Effect of variations in water content of the surface soil (percentage by weight) during summer on the uptake by perennial ryegrass (Lodium perenne) of calcium from the depth of 5 cm, relative to that from deeper soil layers (10, 20 and 40 cm).

to the deeper roots, the restriction of plant growth may be largely due to the limited ability of plants to absorb nutrients in this zone; improved plant growth would thus be expected if nutrients were located at a greater depth in the soil even if water supply remained the same. Garwood and Williams (1967b) illustrated this effect (Table 5.1). Swards of Lolium perenne were subjected to drought regimes similar to those used to obtain the water potential profiles illustrated in Fig. 5.1; the plants absorbed nitrogen from a fertilizer source considerably more readily when it was placed some depth below the soil surface, whereas when the supply of water was abundant superficial applications were effective

It has sometimes been suggested that if a root experiences a gradient of soil water potential it is able to transfer water from the soil of higher to lower potential. Klute and Peters (1969) have suggested that this does not occur to an important extent. The results in Fig. 5.5 and Table 5.1 accord with this view.

TABLE 5.1

Effect of the Depth at which Ammonium Nitrate is Placed in the Soil on the Growth of Perennial Ryegrass (Lolium Perenne) Under drought Conditions.

	Ammoniun nitrate equivalent to 112 kgN ha^{-1} Applied to all treatments		
Depth of placement of fertilizer (cm beneath soil surface)	0	46	76
Increase in Yield of herbage due to fertilizer: per cent*	57	82	102
Apparent recovery of fertilizer nitrogen : per cent**	35	75	80

* Man yield in absence of fertilizer was 2.2 tonne dry matter ha^{-1}

** Apparent recovery of fertilizer nitrogen taken as increase in nitrogen content of herbage due fertilzer.

The Survival of Water Stress

Definition of Water Stress

The term 'water' can convey different meanings depending on whether interest centres mainly on physiological, ecological or agronomic aspects.

In the strict physiological sense plants can be regarded as suffering a degree of water stress when cells are not fully turgid—that is to say when their water potential falls below zero. Mild water stress as thus defined occurs whenever plants are transpiring at an appreciable rate; this is implicit in the fact that the flow of water through plants depends on a gradient of decreasing water potential from the roots to the mesophyll of the leaves. However, in envirnoments favourable for plant growth the continued upward movement of water during the hours of darkness, when transpiration has ceased, largely or completely restores the turgor of leaves.

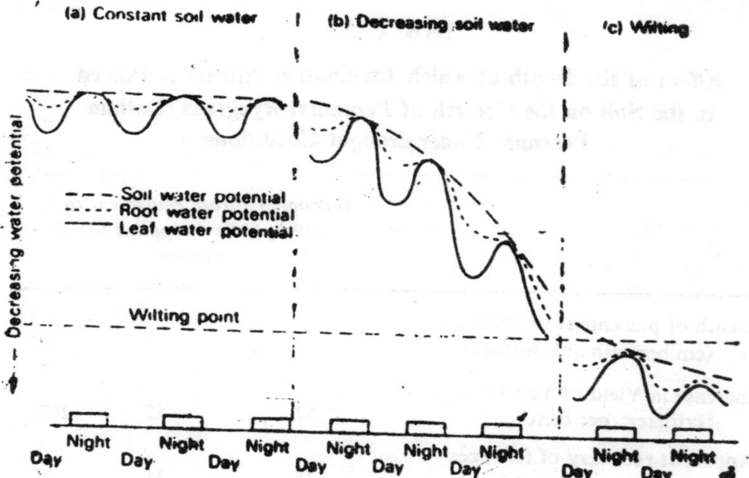

Fig. 5.6: Possible relationships between water potential in soil, root and leaf under conditions of appreciable transpiration when the soil water potential remains high and constant (a), is decreasing (b) and when the wilting point is reached (c)

Such situations, which are represented diagrammatically in Fig. 5.6a, are thus normal; none the less, measurements of leaf water potential during daylight hours indicate an appreciable loss of turgor.

From the ecological or agronomic viewpoint water stress becomes of concern when it imposes a conspicuous restraint on plant growth. In dry weather, when evaportranspiration exceeds precipitation and soil water potential becomes progressively lower, the decrease in the water potential in the plant during the hours of daylight may not be offset by the continuing absorption of water at night. This situation, which is usually accompanied by an appreciable restriction of growth, is illustrated in simplified from in Fig 5.6b. Eventually, if the water potential throughout all or at least the major part of the rooting zone falls to the wilting point growth cases. usually with irreversible injury to foliage (Fig. 5.6c). However, regeneration may still be possible.

Characteristics of Shoots which Influence Survival

This discussion is concerned primarily with root growth in relation to the survival of water stress but if this subject is to be seen in perspective the role of the above-ground tissue must be borne in mind. The shoots of no important crop plants possess to a marked degree the characteristics which enable xerophytic plants to withstand considerable periods of desiccation. Note the less, there can be wide differences between species in the extent to which water loss is restricted by stomatal movement in periods of dry weather (Kramer, 1969), In addition, morphological characteristics can greatly influence the ability of plants to recover after relatively severe desiccation. The regeneration of many pasture grasses after summer drought is, for example, due to the presence of well protected buds near the base of shoots. Even when the water potential in the greater part of the rooting zone has been close to wilting point for many weeks in summer, so that growth ceases, a considerable development of foliage can occur in a few weeks after the soil is returned to field capacity. This is well illustrated by the experiments of

Garwood and Williams (1967a) with perennial ryegrass (Lolium perenne): growth had ceased before the water potential of the soil was reduced to the low values indicated in the right hand side of Fig. 5.1, but rapid growth was resumed within a few weeks of ample water being again provided.

Characteristics of Roots which Influence Survival

The most important characteristic of the root systems of many crop plants, which contributes to their survival in dry conditions, is the ability of root axes to extend sufficiently rapidly that they maintain continuing contact with zones of soil in which the water potential remains adequate under dry conditions. This is a characteristic which causes lucerne (alfalfa, Medicago sativa) to be among the most drought resistant to crop plants. The importance of root depth in determining water uptake under dry conditions is well from the work of Long and French (1967) who compared water withdrawal from the soil in a period of dry summer weather by meadow fescue (Festuca prantensis), a relatively deep rooted grass, and the more shallow rooted timothy (Phleum pratense). Meadow fescue absorbed considerably more water from below 35 cm, and also a greater total quantity, than the shallow rooted species. The difference in water extraction between the two species makes it evident that the resistance to the movement of water through the soil to the root zone (sometimes called the pararhizal resistance) can be much greater than the resistance to its movement through the vascular tissue of roots.

Next to root depth, extensive root branching is often the most important characteristic of root systems which favours the uptake of water. If there is an ample supply of water throughout the rooting zone the size of root systems may be more than ample to supply the needs of the plant (Newman, 1974) and the removal of an appreciable part of the root system can have little effect on the total water uptake (Andrews and Newman, 1968). But the situation is very different in dry conditions when steep gradients of water potential make it possible for water to be absorbed by only a small part of the root system.

When water stress occurs for only a limited period a further characteristic of the root system can be of considerable importance; namely the rapidity with which new roots develop after the water potential in the soil—especially often the surface layers—again becomes favourable. Many gramineae can rapidly develop new nodal or crown roots in these circumstances and this can contribute to their recovery after periods of drought; but it is not necessarily the only reason. Evidence that the older parts of cereal roots, in which a considerable part of the cortex has collapsed, can absorb nutrients from solutions, makes it seem probable that after a root has been surrounded by dry soil for an extended period the absorption of water, as well as nutrients, can be resumed when the water supply again becomes ample provided that the vascular tissue remains intact. This question does not, however, appear to have been studied detail.

To sum up, the combined effect of the depth to which roots penetrate, the extent to which they ramify in zones where water is constantly available and their ability to regenerate when the water supply becomes favourable appear to be the most important morphological characteristics of root systems which enable plants to withstand water stress. But this may not always be so. Passioura (1974) has described laboratory experiments, with simulated drought conditions, in which plants depend on water stored in the soil throughout their entire period of growth. He found that the yield of grain of wheat (Triticum aestivum) could be enhanced when only a single seminal axis was allowed to penetrate the soil. The benefit of this treatment was attributed to a more economical use of water during the earlier phases of growth which resulted from the higher hydraulic resistance of the restricted root system.

VARIABILITY—THE DIFFICULTY OF PREDICTION

The response of root systems to variations in water supply under natural conditions is perhaps the most difficult aspect of the behaviour of plants to predict or to describe in quantitative terms. Uncertainty is caused both by the rapid, and largely unpredictable, changes in water supply which seasonal weather can bring about and the equally rapid response of plants to these changed conditions—including sometimes injury due to anaerobiosis caused by waterlogging. Moreover, it is possible that the closeness of contact between roots and the soil, and hence the resistance to the transfer of water, can be affected.

Figure 5.2 illustrates the impacticability of even describing the form of the root system of a single species in a single soil in a truly representative manner. Because root density normally decreases with depth it is not surprising that an exponential relationship can often account for a large part of the variation in root distribution with depth in soil (Gerwitz and Page. 1974). But this is not a constant relationship, it happened in this first (wetter) year illustrated in Fig. 5.2 but in the second (dry) year the relationship was closely linear. In both years, however, there were considerable discrepancies between the density of root observed at each depth and that calculated from the exponential or linear equations which fitted the observations most closely. Mathematical relationships to describe the growth of plant roots in uniform environments have been developed (*e.g.*, Hackett and Rose, 1973; Lungley, 1973). However, these models can at present help little in the study of field problems since an understanding of root function is of the greatest interest when they are subject to stresses—often transient and unpredictable ones. The recognition of this in no way disputes the desirability of seeking as fully as possible to derive quantitative relationships which may assist in describing the response of plants to their environment.

6

Terrestrial Ecosystems

To understand the biosphere, we must examine major biomes and ask not just how communities differ but why. Although no single factor defines an environment, environmental extremes may dramatically limit options. Consider, for example, the earth's poles. These ice-covered regions that the penguins and polar bears call home are the least hospitable areas of the earth. They are cold most of the year because of the pattern of sunlight, water currents, and wind. The brief summers are never warm enough to melt the accumulated ice packs. As a result, rooted plant life is impossible. In the north the food web is based on the producers of the Arctic Ocean. The presence of a continent at the South Pole results in an environment with little life. A well-developed food chain is present only on the edge of the continent.

In the arctic north, the land of permanent ice is edged by a tundra that develops where more moderate day lengths and increased solar radiation leave the soil bare and soft for a few short months. Conditions permit terrestrial food chains to develop. Large, nomadic animals, such as reindeer and caribou, graze on the low, hardy vegetation but must range widely to find enough food. Predatory wolves and foxes follow the great herds or hunt lemmings and other small rodents that burrow in the ground. Although tundra soil is frozen much of the time, sunlight thaws the surface each spring. Suddenly plants burst into flower as the desolation of the dark, snowy winter is broken by the long, warm days of arctic summer. Black flies, mosquitoes, and other insects emerge in great swarms. Migratory waterfowl, shore birds, and song birds arrive from thousands of miles away to nest and feed on hordes of insects from temporary ponds and thawing rivers.

Father south, cold is not so pervading; here the land form and rainfall become important variables. As a result, the earth is blanketed with biomes like a giant patchwork quilt. Forests, grasslands, and deserts are scattered about. No two are alike, and of course, none remains untouched by human hands. Some we have only exploited. Others we have manipulated beyond all recognition. The story is at the same time fascinating, inspiring, and sad.

TEMPERATE DECIDUOUS FORESTS

European colonists found most of the eastern seaboard of North America covered with deciduous trees—trees that shed their broad leaves in the fall. Similar forests once blanketed most of Europe, the temperate coasts of Asia, and smaller areas of Australia and South America.

The Plant Community

Mature deciduous forests often contain dozens of tree species, but in any given area one or two kinds of oak, hickory, chestnut, maple, basswood, buckeye, or beech are common enough to give the forest their name. Big trees dominate the community by providing most of the food supply and by modifying the physical environment through shade and windbreak action. The largest trees are widely spaced, but the forest includes individuals of all ages. Such a forest is a stable community that can maintain itself through continued reproduction.

Under the trees of a deciduous forest, shrubs and herbaceous (nonwoody) plants are scattered about, but few are abundant except in clearings. The density of the leafy canopy overhead and the time each year that the canopy is present regulate growth of understory (below the canopy) plants. Many of these are small herbs or "spring flowers" that grow rapidly as days lengthen. They reproduce and carry out much of their photosysnthesis before the tree leaves open and reduce the sunlight available on the forest floor.

Soils of some deciduous temperate forests are brown and rich-looking, but most of the available minerals cycle through the plants each year. Only a small portion of the minerals absorbed are retained in the wood of the tree. The remainder returns to the ground as twig and leaf litter. Not all the returned materials can be reused the following year, because decomposition takes a long time. Of the minerals released from the litter, some invariably leach away in rainwater. Humus from decomposing leaves and twigs helps hold both water and minerals.

Temperate deciduous forests are highly productive, and they support a large number of organisms, as suggested by Table 6.1. Such forests are also quite diverse. An often overlooked part of the community lives in the litter on the forest floor. Much of the forest enters this decomposer food chain yearly. Litter decomposition involves the activity of multitudes of tiny animals and microorganisms. Insects such as springtails and other near-microscopic arthropods eat dead leaves. These primary consumers absorb less than 10 per cent of the nutrients in the litter. However, as a result of their digestive process, the remains that become feces are easily attacked by bacteria and fungi.

Animals not only consume and alter litter; they also mix it into the soil. Earthwarms are among the most active litter mixers. The carnivores that eat the litter feeders also stir litter and mix it into the soil as they endlessly seek their prey.

TABLE 6.1

Average Numbers of Organisms/2.6 km² (I Square Mile) of Temperate North American Deciduous Forest During the Summer*

Plants	
Trees 7.5 cm (3 in.) or more in diameter	75,000
Tree seedlings	78,600
Shrubs	281,000
Soft-stemmed plants	345,000,000
Animals	
Invertebrates (insects, snails, centipedes, millipedes, earthworms, etc.)	2,688,000,000
Pairs of small nesting birds	768
Large predatory birds (owls and hawks)	2 to 5
Mice	240,000
Gray squirrels	1,500
Flying squirrels	1,500
White-tailed deer	40
Wild turkeys	20
Gray fox	3
Black bear	0.5
Mountain lions	0.2

* From estimates compiled by Victor E. Shelford in The Ecology of North America, University of Illinois Press, Urbana, 1963.

Adaptation to Climate

Temperate deciduous forest have moderate temperatures and rainfall (about 70-100 cm or 28-40 in./yr) but are characteristically subject to forest. Shedding leaves in the fall protects the trees against cold damage. Otherwise the large leaf surface that favours transpiration and photosynthesis during the warm season would allow heavy evaporation and severe water loss during the winter. You will recall that the absorption of water by roots relies on diffusion, as well as on energy-consuming activities of root cells. Both are slowed by cold, and of course, frozen soil yields nc water. Cold weather can dehydrate plants more than hot weather, especially when it is combined with strong winds. Furthermore, leaves are difficult to protect against freezing. Deciduous trees have evolved an effective strategy to prevent damage from winter cold. First they concentrate sugars and other organic compounds from their leaves into roots and stems. Here these large molecules lower the freezing point of cell fluids and prevent formation of ice crystals that could rupture cells. Then the trees simply shed their leaves.

Leaf fall is a preprogrammed process coded into the genetic structure of the plant and only slightly affected by the weather any particular year. In response to a decrease in auxins, an abscission (cutting off) layer forms where the leaf stalk joins the stem (Fig. 6.1), and the cement between the cells in that area softens. In the absence of strong intercellular cements, the short cells of the abscission layer separate, permitting the leaf to drop off or blow away in a gentle breeze.

The colour changes that foretell leaf drop in deciduous forests result from cessation of chlorophyll formation. Because chlorophyll breaks down spontaneously in light, leaves remain green only through continued synthesis of chlorophyll. With the coming of fall, chlorophyll manufacture ceases, and this green pigment gradually bleaches in sunlight. In the absence of chlorophyll we see yellow or reddish pigments, mainly carotenoids that aid in light absorption for photosynthesis. Another source of red colour results from interruptian of phloem by the developing abscission layer. This causes sugars to accumulate in the leaves, where some are converted into reddish compounds.

Community Succession in Deciduous Forests

When the vegetation in an area is destroyed, it is usually

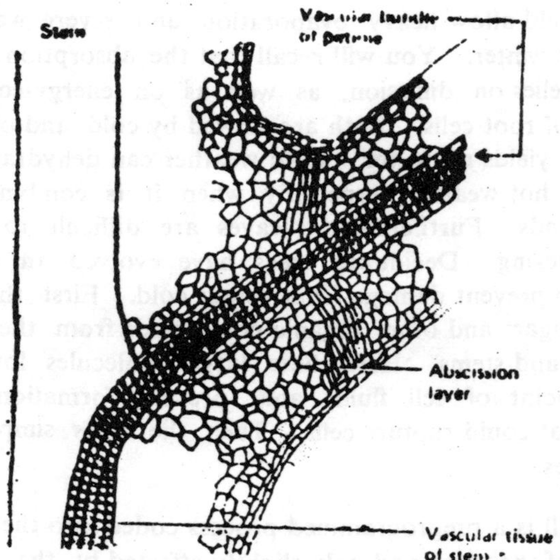

Fig. 6.1: **Abscission Layer Aids Leaf Drop. The abscission layer is a band of specialized cells in the petiole at the base of a leaf. Separation here leads to natural leat fall. The shedding of leaves is initiated by the plugging of the vascular tissue in the abscission layer. As a result, the leaf tissues gradually die. Formation of short cells within the abscission layer and softening of cement between such cells permits the leaf to break away cleanly. By the time this happens, the abscission cells have already healed over the leaf scar and protected the tree from invading organisms**

replaced. Often the new vegetation is different from what was dertroyed. And in most cases, this new vegetation is gradually replaced with another and then another. The sequence for any particular area is always the same. Eventually the vegetation becomes stabilized. This permanent, stable vegetation is said to be the climax vegetation for that particular region. The replacement of one vegetation, and then another, in a predictable sequence is known as succession. The fact that successional stages and climax vegetations differ from place to place results from differences both in the environment and in the history of the vegetation present. Obviously, a particular species can appear in a successional vegetation only if there is a seed source.

The seed may be present in the ground or provided by patches of successional vegetation. Because climax vegetation is susceptible to destruction by fire, flood, or some other catastrophe, no area is long without successional growth. Most human societies continually disrupt natural vegetation. Consequently, the vegetation people see daily is successional rather than climax vegetation. Examples may be found in roadsides, fence rows, neglected yards and fields, and cut-over land.

When first visited by Europeans, much of the Americas supported climax vegetations. Because the forests seemed untouched by humans, they came to be known as virgin forests. Usually what the layperson calls a virgin forest the biologist recognizes as the local climax vegetation. The common term second growth most often refers to a successional forest.

The factors controlling successional patterns differ from place to place, and they are often poorly understood. Most studies have involved land that was cleared for agricultural use and then, later, abandoned. Whenever bare ground is left undisturbed. rapidly growing annuals (plants that set seed the first year and then die) may appear. Although there is a characteristic first-year vegetation in each plant community, in many places these plants are replaced by introduced annuals that everyone calls "weeds." This is one example of the fact that the history of an area affects the pattern of succession.

During the first year, biennial (two-year) and perennial (many-year) plants also get started. The second spring these plants draw on reserves, usually stored in their roots, and make such rapid growth that they crowd out the young annuals. Many biennial species have their early leaves in a whorl close against the ground. These leaves create dense shade around the base of the plant and contribute to suppression of nearby annuals. The dandelion is a biennial with this characteristic grawth pattern.

Although the details vary for one reason or another, one stage of succession leads to another. The growth habits of annuals and biennials may explain which are dominant the first and second year in certain biomes, but many other factors can be involved. Shade tolerance at various stages in the life cycle differs among tree species and may determine the pattern of forest succession. In much of the deciduous forest biome, pines precede the climax vegetation. Most pines are particularly vulnerable to shading. Lower limbs die as the pine forest grows dense, and pine seedlings never survive in deep forest shade. The onks, maples. beeches, and other members of the climax forest are more shade-tolerant. Small trees become tell and thin in heavy shade but survive and eventually grow taller than the pines. Soon shade from the deciduous species kills the pines. The seedlings of the deciduous trees also survive in the shade of larger ones, so there are always small trees in the understory. When a tree is blown over, others quickly replace it in the canopy. Thus shade tolerance helps explain why the deciduous trees replace the pines and also why the deciduous trees are the climax vegetation. For details of succession in a well-studied part of the deciduous forest biome.

The series of vegetational stages that follow destruction of a climax vegetation, as we have just described, is often called secondary succession. Note that secondary succession begins when soil is available. Primary succession is a series of events whereby plants colonize an area of barren rock, such as fresh lava. Presumably this process has occurred at least once wherever there is now vegetation. In many instances primary successions have been associated with the evolution of species and have required millions of years. However, we see examples of the early stages of primary succession today on rock outcrops. The original colonizing plants are usually lichehs that produce acids. The acids dissolve certain minerals in the rocks. This process roughens the surface, so the lichens can attach better, and it also makes minerals available for lichen growth. Over the years, mineral grains released by the acids and decaying lichen components create a small amount of soil. Other plants, particularly mosses, may grow in the lichen soil and so constitute the second stage of primary succession. Later events depend on the environment and the species present, just as we noted with secondary succession.

What Humans Have Done with Deciduous Woodlands

Most of Europe was once covered with climax deciduous forests except at high elevations and in places where standing water created marshes. The forest canopy was so dense that, it has been said, a squirrel could travel from central Russia to the Atlantic coast without ever touching ground. On much of the continent the dominant species were oaks, but in some regions elms, beeches, and birches were prominent.

Fire and Axe. Many of these forests succumbed to fire or the flint axe. At first people cut trees mainly to obtain building material or fuel, but sometimes trees were felled merely to clear the land. Initial clearings were often abandoned and recovered by forest, as has been shown by the study of pollen buried in the mud of nearby lakes and bogs. Decreases in pollen of oak ivy (a native understory vine) mark the first clearing of the forest. This stage is followed by high levels of weed pollen, which indicate a period of cultivation. Later peaks in hazel pollen show when these shrubby trees, took over fields that must have been abandoned.

Vast tracts of climax forest still flourished in Roman times. Throughout that period the forest edge expanded and contracted with wars and other changes in human fortunes. Eventually, the relentless demand for plowed land and forest products doomed the woodlands. As commerce and trade grew, more and more timber was needed for shipbuilding. Glass and soap manufacture required wood ashes as raw materials. The smelting of tin, lead, copper, and iron depended on charcoal made from wood. By the eighteenth century these industries had scalped the forests of England, and the British had turned to digging coal. The woodlands of western Europe disappeared a short time later.

Such drastic alteration of the landscape naturally had far-reaching effects. Rains began to wash the soil from the naked land. It is no coincidence that several British seaports filled with silt during the twelfth and thirteenth centuries and were abandoned when their waters were no longer deep enough to float seagoing vessels.

Erosion is still evident in parts of England. Some fields have been farmed with the same hedgerow boundaries ever since the land was cleared. In places where the hedges run across the middle of slopes, the soil is as deep as six feet on the upward side of the hedgerow but so thin on the lower side that rocks lie exposed. We have no idea how much more soil eroded away entirely and washed into the ocean.

Domestic animals prevented regeneration of most British woodlands. A great sheep industry thrived in England during the Middle Ages; grazing was so heavy that the seedlings had no opportunity to grow out of reach of livestock. As sheep-rearing decreased, rabbits, introduced from the Continent, increased dramatically. The rabbits stripped bark from the young trees and did so much damage that few seedlings survived.

Conversion to Grassland and Moor. In many deforested areas of Britain the climate, soil, and grazing animals interacted to create a habitat favourable to coarse grasses. Manure from the sheep and rabbits returned minerals to the soil, established an efficient nutrient cycle, and permitted the formation of a crumbly soil that supported a heavy turf.

Not all the deforested British lands went to grass. On some soils the loss of trees raised the underground water level enough to create bogs. There is a simple explanation for this drastic change. Each day during the leafy season, large trees transpire tremendous volumes of water. Deforestation reduces this drain on the groundwater. Therefore cutting certain forest can lead to their replacement with bogs or wet, peaty moors that support low shrubs.

Fate of the Forests. Wherever civilization flourishes, forests are destroyed. The history of other forests differs little from that of European forests. The Chinese began clearing ground for crops at least four thousand years ago. The Mediterranean forests were already seriously damaged by Plato's times. As a rule, few forests remain where civilization has long thrived.

Although we must infer much of the history of Eurasian vegetation, we know a great deal about what happened in North America. Along the East Coast, the abundance of land and the need for a lightweight crop the could be traded in Europe promoted slash-and-burn tobacco farming. Not only did cutting and burning the trees clear the land, but it provided ashes that contained sufficient minerals to support a few crops of tobacco. However, these minerals were soon lost from American soil. The mineral-rich tobacco leaves were shipped to Europe and connerted again to ashes, this time in foreign pipes. The broken nutrient cycles and rapid erosion quickly ruined the tobacco plantations. Whenever that happened, the owners merely moved westward and cut another area. In this way the forests around Chesapeake Bay were destroyed before the American Revolution. So much soil washed into streams and rivers and settled into the bay that the shoreline we know is quite different from the one the early settlers found.

Wherever the tobacco fields, were abandoned, the vegetation began to regenerate. Probably some fields in Virginia and nearby states have been cleared three or four times over as many centuries. A few bear climax vegetation today, but because of the chestnut blight, we have not a single forest just like those the first planters cut down. Chestnut blight is a fungus introduced into New York early in this century on disease-resistant Chinese chestnuts. Soon the blight swept through the eastern part of the United States leaving great stands of dead trees. Some old chastnuts still have live roots that send up new shoots, but there are no new trees, because the blight is lethal to sprouts.

CONIFEROUS FORESTS

The conifers are gymnosperms that bear seeds in cones instead of within fruits as do the flowering plants. Familiar conifers include pines, spruces, firs, cedars, and redwoods.

Boreal Forests

A wide band of conifers girdles Eurasia and North America north of the deciduous forests and grasslands. Most of these conifers retain their leaves the year around and shed one set only after the next is in place. Persistence of the leaves permits conifers to take immediate advantage of good weather, since they are able to begin photosynthesis without waiting to develop new leaves. This is a distinct advantage to ward the poles, where the warm season is short.

These conifers also tolerate low temperatures. One reason is that they have stiff, wax-covered leaves known as "needles." When cold winds blow through the forests, the thick layer of wax retards water loss through evaporation. Of course, the leaves do dehydrate, and they would wilt if they were not rigid. Thus the stiffness of the needles, due to tissues with exceedingly thick cell walls, prevents damage that would otherwise be caused by wilting. The leaf structure of conifers clearly suits them to subarctic regions, where winter winds sweep across the landscape.

Coniferous communities have fewer species than do deciduous forests. One reason is that the ever-present shade of the year-round canopy restricts photosynthesis by understory plants. As a result, the ground is nearly bare except for a deep carpet of slowly decomposing needles. Slow decomposition coupled with leaching of the soil by acids from the needle litter results in low fertility. Conifers generally have root fungi, known as mycorrhizae, which aid in accumulation of minerals. Trees without these symbiotic partners are at a disadvantage.

A boreal forest usually consists of one species of high latitudes. The major differences are in the length of the day and the intensity of sunlight. Where mountains are at temperate latitudes, organisms experience a day length typical of the temperate zone. But because mountains stand above much of the earth's atmosphare, the sun's radiation is more intense here than at lower attitudes. (This results from the fact that the atmosphere absorbs part of the radiation that reaches it.) Yet mountains cool quickly at night because of loss of radiant heat through the thin atmosphere. This and exposure to winds make high altitudes colder than those in adjacent lowlands. The cool weather of high altitudes creats environments resembling those toward the poles.

Southern Conifers: A Fire Climax

The same adaptations that fit conifers to winter dryness in cold climates suit many to hotter areas, especially where the soil is sandy and the water supply is unstable. In the southeastern United States, pines colonize abandoned fields and form major forests. Perhaps their initial success depends on mycorrhizae, since these soils are usually poor. Left undisturbed, many such pine forests are replaced by oak, hickory, or magnolia. Apparently fire set by lightning or by humans maintains these pine forests, since they tolerate fire, whereas the broad-leafed trees do not. The pines have a thin trunk bark through which new buds sprout if existing limbs are destroyed. A thick tuft of needles protects the terminal bud of the southern longleaf pine and makes it extremely resistant to fire. As a matter of fact, longleaf pine seedlings seldom survive in the absence of fire, for they are sensitive to crowding and shading. Only when other vegetation is burned back can the little pines get enough light to grow.

If southern pine forests are burned every few years, the fires do little damage to wildlife, because the animals can escape by running through of flying over the low fire line. In fact, occasional burnings of patches of pine forest increase the population of bob-white quail and wild turkeys. These birds find food and shelter in the successional stages that follow the fires. We are not suggesting that fire is always beneficial even in these forests. In contrast to the low, cool surface fires of frequently burned regions, fires in pine forests that have not burned recently blaze up high and hot. Over many years without fire, a thick blanket of needles and fallen limbs accumulates on the ground. Once this litter is on fire, it provides so much fuel that the heat is sufficient to ignite the tree-tops and produce great crown fires. Such fires destroy the plant community and much of the wildlife.

GRASSLANDS

Most temperate grasslands have long since succumbed to the cow and the plow. Only a few retain the unmodified vegetation of the plains, prairies, steppes, velds, or pampas. But whether they bear, native vegetation or support the highly selected grasses we know as grains, these grasslands supply the bulk of human food. Cattle and sheep graze on their grasses; cattle, hogs, and chickens fatten on corn grown on former grasslands; and most of our cereals, especially wheat, are produced on soils that once supported native grasses.

Biology of Grasses

All the grasses belong to one large family and share distinctive characteristics that fit them to their environment and make them of particular use to humans. Many people are surprised to learn that grasses are flowering plants. Open grasslands are windy, and it is the wind that pollinates grass flowers. Among these wind-pollinated plants, natural selection has favoured those that have conserved resources by producing small, petal-less flowers. Therefore, grasses and some other wind-pollinated species lack the showy petals that other plants use to attract animal pollinators.

A single head of oats or other grass contains numerous flower. Each tiny flower can form a single fruit called a grain, or kernel. The overian tissue of the fruit fuses with the single seed inside in such a way that they become one continuous structure.

Adapted to Withstand Grazing. As grasslands evolved, their abundant thin leaves offered a bonanza to herbivorous animals. As a result, there evolved a large group of grass-eating mammals, including the ancestors of today's cattle, sheep, and horses. The selective pressures of these grazers promoted evolution of traits that resist grazing damage. For one thing, grasses store nutrients in their roots. As a result, loss of stems does not substantially reduce their reserves. Another factor is the presence of silica, the substance we know as glass, in grass cell walls. Of course, grazers have evolved excellent grinding teeth. But grasses take their toll. Tooth wear is so severe that many old herbivores eventually starve to death.

The growth pattern of grasses has several distinctive characteristics that many also adapt them to withstand grazing. Grass leaves originate singly at prominent nodes along the stem. The lower part of each leaf forms a sheath that surrounds the stem for considerable distance before bending sharply outward as a flattened blade. The leaves retain a meristematic growth region at the base of the sheath and another at the base of the blade. If the blade is cropped back, these meristems resume growth and lengthen the leaf to compensate for lost photosynthetic tissue. Thus grass leaves grow from the base rather than from the tip. You can see this growth pattern in any lawn.

Until time to flower, the grass stem remains short, and the leaves grow upward beyond the tip of the stem. Thus the stem tip is protected from damage as long as possible. Only when it is time to reproduce does the stem elongate between several nodes and the tip stretch skyward. Here the flowers develop where they are exposed to the winds that pollinate them.

Binding the Soil. Grasses bear abundant roots. Most are adventitious roots, which arise not from the primary root formed by the embryo but from the lower nodes of the stem. The sod-forming perennial grasses have extensive stems lying just on top of the soil or in the upper soil layer; these serve both to hold the soil and to spread the plant. From rhizomes (underground stems) arise adventitious roots and occasional branches that penetrate to the surface, giving rise to what appear to be new plants. Quackgrass, a garden pest across the northern half of the United States, often reproduces through rhizomes. Plowing or cultivating disrupts and scatters the quackgrass rhizomes. Each piece with an intact node may form a new plant. Often grasses have stolons, stems that lie on the ground and root at intervals. Each rooting node may give rise to aerial branches.

Rhizomes and fibrous roots constitute over half the mass of a grass plant. Together they form a network penetrating throughout the soil and binding it into a nearly inseparable plant-soil complex known as sod. Neither wind nor water can erode a healthy sod. Growing roots and rhizomes break the soil repeatedly and contribute to the characteristic crumbly texture of grassland soils. Many of the roots of perennials die each year and are replaced by new roots. Decay of dead roots supplies humus and leaves speces that aerate the soil. Decaying roots return minerals to the soil, where they are immediately reclaimed by other roots. This process is part of the rich and efficient nutrient cycle of grasslands.

Dead stems and leaves provide an extensive ground cover in most grasslands. Unmowed grasses accumulate as much as 10,000 kg of humus per hectare (about 9000 lb/acre) each year. It takes three of four years for litter components to decompose. Hence the litter forms a deep layer that shades the ground and lessens evaporation due to the wind. The litter also holds rainwater and aids its penetration into the soil. Because the water soaks in, there is no surface runoff and therefore little erosion. Except in the most moist grasslands, the rainwater selom soaks deep enough to join the groundwater. Instead, the extensive root system picks up most rainfall while it lies in the upper layer of the soil. Then the xylem carries the water immediately to the leaves, where it is transpired back into the atmosphere. Because little water percolates down far into grassland soils, minerals are not leached away; instead, they remain near the roots. Retention of minerals contributes to the richness of grassland soils. This and the crumbly texture produced by the roots make grassland soils excellent for agriculture.

Grassland Communities

The factors that limit forests and permit grasslands to develop are hard to determine. Some biologists maintain that grasslands develop only where there is insufficient rainfall to support trees that could shade out the grasses. On the other hand, there is both historical and experimental evidence that many grassland edges, such as those between the American prairies and the eastern deciduous forest, were maintained by fire. Where forests and grasslands meet, burning usually favours grasses over trees.

Variations in climate and accidents of grazing have resulted in complex and dynamically balanced grassland communities. Contrary to popular belief, the American prairies were never a uniform sea of grass ranging from the deciduous forests of the East westward to the Rocky Mountains. Mixed in with the grasses there were other herbs, especially members of the aster family and legumes, such as the lupines. Nitrogen fixation by symbiotic bacteria in the nodules of the legume roots contributes to the fertility of the soil and the vigour of the community. The roots of various species extended to different levels and were best developed in specific regions of the soil. Concentration of roots of one species at a particular level reduces competition between species. But because each level is populated by roots of some type, resources are fully exploited.

The grassland landscape is even more varied along the moist borders of streams, rivers, marshes, lakes, and ponds. Where inland waterway pass through grasslands, groves of trees and shrubs thrive, along with other organisms ordinarily associated with woodland communities.

Grassland Animals. A large and diverse animal community lives in every grassland. The deep soil that covers rocks and the absence of woody vegetation limit aboveground shelter, except where grasslands merge with woodland or along rivers and streams. Thus most small animals depend on the soil for protection. Numerous little mammals burrow beneath the sod.

Prairie dogs, rodents related to squirrels, are symbolic of the short grasslands or Great Plains of North America. Stockmen exterminated most of the prairie dogs, believing that these animals competed with livestock for forage. Undoubtedly the selective grazing of prairie dogs on grasses favoured the growth of other plants. Pronghorn antelopes, which were once almost as numerous as bison and are now few in numbers, relied heavily on the broad-leaved vegetation of prairie dog "towns." Elimination of prairie dogs also resulted in near extinction of one of its predators, the black-footed ferret. This handsome, weasellike animal hunted in the prairie dog burrows. Because vast numbers of prairie dogs are necessary to support even a tiny breeding population of ferrets, this species is now extremely rare—or perhaps extinct except for a few in captivity.

Prairie dog burrows and those of gophers, pocket mice, kangaroo rats, and ground squirrels provide shelter for other animals, including grass-hopper-mice and snakes. At least one bird, the burrowing owl, relies on old rodent burrows for nesting sites. So do cottontail rabbits. Beetles and camel crickets also use burrows; the dung, fungi, and hoarded vegetation afford these insects a ready food supply. Ants are abundant in grassland soils; some species build huge mounds surrounded by a zone stripped clean of all visible life except for the ants themselves. Animals that dig underground benefit the grasses, because they aerate the soil and mix in humus from the surface. Even those that are underground transients contribute fecal material where it is directly available to the roots.

Some animals, such as grasshoppers, are found only aboveground. Nevertheless, they depend on the soil for shelter. The female grasshopper uses the tip of her abdomen to penetrate deep into the soil and bury her eggs. The adults perish in the rigour of winter, but the species continues because the developing embryos of the next generation lie safely protected in the soil.

A number of medium-sized animals that live in the surface vegetation face danger from predators. Rapid movement through thick grasses is very difficult, and vision is limited; consequently, it is easy for predators to stalk their prey. Escape must be fast and sure; otherwise it comes too late. Consider how grasshoppers, jackrabbits, and jumping mice meet this challenge. Huge aerial leaps permit such animals to clear the top of the vegetation, get an unimpeded view for a moment, and then drop some distance away without leaving a trail.

No Hiding Place. In wide-open habitats large animals find little or on shelter. Under these circumstance "predator control" becomes as group activity, and the animals aggregate together. The pronghorns, bison, and muskoxen of North America, the kangaroos of Australia, the saga antelopes, wild horses, and asses of the steppes of Russia, the gnu and zebra and even the ostriches of Africa show herd instincts. In a group there are many eyes to watch for predators, and alarm spreads rapidly. Frightened pronghorns raise their tails and display a large white patch on their rumps. The fiash of white rump and white flaglike tail on a running pronghorn starts every other pronghorn in sight on its way! Mixed herds, such as are common in Africa, cooperate in watching for danger. Ostriches are taller than most mammals they herd with and are usually the first to sound alarm.

Large grassland animals are exceedingly fleet; they have long leng that cover the ground quickly. The faster runners in the world inhabit grasslands. The American pronghorn, our fastest native animal, can do 60 miles an hour. At this rate the pronghorns leave a solitary predator or a pack of wolves so far behind that no amount of strategy or cunning can lead to another encounter until the pronghorns have had time to feed and recuperate from the original confrontation. Of course, natural selection has also promoted evolution of fast-running grassland predators. In Africa, the cheetah has been clocked at 65 miles an hour. It is known to accelerate from a stand-still to 45 miles an hour in a few seconds.

Grassland herbivores rely on the group not only for detection of predators but also for defense. When threatend, bison form protective head-outward circles. The young are secure in the center of the circle.

Herding is itself a defensive measure. Predators are confused by large numbers of running animals. Thus if the individual a predator is following loses itself within the herd, the predator stops in confusion. For this reason, most predators kill only aged, diseased, deformed, and very young animals that are unable to keep up with the herd. These same animals are especially vulnerable, because they are also weak.

Grazing and Overgrazing

Each grassland is a balanced ecosystem of producers and consumers. The native herbivoures have evolved with the plants and are adapted to the vegetation and to one another. Likewise. grasslands are adapted to withstand grazing by certain animals. Unfortunately, domestic animals often have destructive effects on ecosystems in which they did not evolve.

The abuse of natural grasslands by domestic animals arises from two factors. First, the domesticated grazers, with few exceptions, are from other parts of the world. Hence local grasses are not adapted to circumvent their feeding patterns. Domestic sheep, in parlicular, can graze more closely than native North American herbivoures can. Most grasses are vulnerable to extensive loss of stems and leaves.

Second, overgrazing by domestic animals is common. The usual population controls that limit wild grazers are nearly if not totally absent from populations of domestic animals. Humans do their best to protect their animals from severe weather, predators, or temporary food shortages. When they slaughter animals for food, people carefully select surplus young adults, particularly excess males. Very young animals, which have a period of rapid growth ahead of them, pregnant females, and females of good breeding age are seldom killed. In almost all cultures, people exploit their understanding of animal growth and reproduction to maximize the number of animals and obtain as much food as possible. Unfortunately, most societies lack a similar understanding of the ability of grasslands to support animals for long period. Perhaps this difference in understanding animals compared with grasslands results from the fact that the life cycle of the animals is comparatively short. Each human has the opportunity to see many generations of cattle grow up and reproduce. In comparison, the effects of overgrazing may appear only over decades and sometimes require several human generations to become acute.

Damage from Overgrazing. Repeated close cropping of grass blades reduces the photosynthetic potential of the plant and hence the energy that can be stored in the roots to support the following seasons' growth. In response to grazing, most grasses undertake a compensatory growth. The leaves elongate at their meristems, and new branch stems arise near the ground. Both results from the loss of auxin-producing tissue at the tips o f leaves and stems. Under excessive grazing some grasses continue to make new growth until their underground stores are exhausted.

Heavy grazing prevents the stems from growing out, blooming, and setting seed. Sometime the supply of new plants is cut off completely. Close cropping of leaves and stems also inhibits normal growth of roots that must occur to replace old, dying roots, to ensure absorption of water and minerals, and to bind the soil against erosion. If all leaves are gone, the grass crown (base of the stem) lies exposed to physical damage from freezing or trampling.

Finally, overgrazing harms hams the soil directly. Reduced vegetation and surface litter increase evaporation and reduce water absorption. Much of the rain simply runs off along the surface, eroding the soil. Overgrazing has accelerated erosion of the deep gullies and arroyos of the arid Southwest. Heavy grazing also changes the soil texture. Continuous trampling compacts barren soil and thereby destroys spaces between the particles. This reduces ability of the soil to hold air and water.

Range Management. Except for the constant pressure for short-term economic gains, range mangement is quite simple. Grasslands must be permitteded to set seed. The manager can achieve this goal by dividing the land into plots and rotating the grazing so that each plot remains ungrazed every few seasons until after the seed has ripened and dipersed. If overgrazing on the other plots is to be prevented, more land must be available for a herd of a given size, or the total number of animals must be reduced. Another problem arises from the fact that grazing animals never use available forage uniformly. They always prefer some areas, especially those near water holes. Sometimes judicious placement of salt blocks can balance the distribution of grazing.

Natural Checks. If range use must be so carefully managed, how did the thundering herds of bison live in equilibrium with the native grasslands of North America? The answer is quite simple: there really were not many bison in relation to the amount of grassland. The highest estimates of bison populations suggest that there was only one animal per 20 acres of range. Although the bison traveled in herds, raising clouds of dust with their hoofs and trampling the vegetation badly in places, most plants went many seasons without a single bison passing their way. Some pond and stream margins did suffer repeated damage, and these probably supported successional vegetation. But in general the grasslands were lightly used. Natural checks limited the bison population. Although only grizzlies from the western mountains could have killed healthy adults, the young bison and ill ones were prey for wolves and perhaps coyotes.

Weather may have been another limiting factor. According to Indian tradition, a large herd of bison disappeared from Illinois during a terrible blizzard late in the eighteenth century. Always some bison became trapped in river muds, and some surely died from disease or parasites. But the best animals usually survived, and numbers were in balance with the food supply. However, there can be little doubt that there were times when drought brought starvation to the bison and damage to the grasslands.

Drought, Dust, and Deserts

The moderately dry climate ot grasslands carries the constant risk that fluctuation in rainfall will create drought. Twenty-three separate droughts have been recorded in the Russian steppes alone during the past century and a half. But the geassland community survives, because the plants are adapted to the ravages of drought. Overgrazing reduces the resistance of grasses to drought, and plowing the sod can lead to erosion that will nearly destroy the entire ecosystem. Let us examine the history of American grasslands and the effect of humans and drought on them.

The Dust Bowl. The late 1920s and early 1930s were marked with increased rainfall that permitted tall grasses to to flourish on the American prairie. The seven-year drought that struck in 1933 drastically altered the species composition of the unplowed prairies, most of which had been subjected to heavy grazing. Dry-adapted grasses quickly replaced those requiring more moisture. The big bluestem grass that could grow nearly eight feet high almost disappeared ; only a deep root system and underground food reserves permitted some plants to survive.

Even more drastic changes occurred in the normally drier Great Plains to the west. Grazed ares were badly damaged, but ranchers were reluctant to reduce their herds. One study in western Kansas showed less than half of the ground covered with vegetation in 1935 and only five per cent in 1936. Great dust storms were both the cause and the result of the decreased vegetation. High temperatures and low rainfall prevented crop growth on plowed fields. Dry winds raised clousd of dust from these fields and from the poorly covered grassland; when the dust fell from the air, it drifted like a horrible dark snow. A deposit of only an inch was often sufficient to smother short gasses and produce another field of unstable soil.

Stressful as the great drought was, no species is known to have been completely eradicated. All survived in favourable habitats or as ungerminated seeds. Such survival would be expected, for only a small portion of the seeds of wild plants will sprout upon encountering favourable conditions. Nor will all the remainder sprout in any one season. So after several years in which all germinating grass seedlings shrivel from lack of water and die, there will still be live seed left in the ground. When a normal year finally comes, a few seeds will sprout and begin to replenish their kind.

Although no species were lost in the drought, the effect on the soil was another story. The rains that followed the drought fell on loose, naked soil that eroded into steep gullies.

Turning Grasslands into Deserts. The drought of the 1930s was certainly not the first one Americans had known, nor was that the first time dry weather and overgrazing had damaged the grasslands and hurt the ranchers. Increasing herds of sheep and cattle during the latter part of the nineteenth century had culminated in starvation of stock and disastrous economic losses when drought struck in the 1890s.

Following the Spanish pattern of migratory sheep grazing, some regions of the West have been overgrazed for several centuries. By moving the animals continuously, mainly to higher elevations in the spring and back downwards in the fall, ranchers can maintain large herds on very poor land. Arid western lands often show the effect of this practice. The hills are terraced by continuous rows of sheep paths. Except on th highest peaks of the Sierra Nevada and the Rockies, the vegetation has been altered, perhaps permanently, by such practices.

In some regions natural grasslands have been replaced by weedy desert vegetation. In western North America sagebrush now dominates many of what were once cool, short grasslands. To the south, creosote bush, cacti, and mosquite have replaced other grasses. It is important to recognize that these "new deserts" lack the diversity and complexity of old "true deserts."

But once overgrazing has produced bare sports, the mesquite thrives. It sends roots out literally, as well as deep into the soil, sometimes as far as 8 meters (25 feet). Overgrazed grasses with poor root systems cannot compete for water with mesquite. Soon the grasses die back further, and the mesquite expands. Hungry cattle feed on the fleshy mesquite fruit pods, but mature seeds pass unharmed through the cattle. Thus cattle scatter mesquite seeds about, especially in the trampled areas where they congregate. Fertilized by the cattle dung, the mesquite seeds are well planted in the barren but hospitable soil.

American cattlemen have no monopoly on the conversion of grasslands to deserts. It has long been said that the Sahara Deserts marches farther south each year. During the early 1970s drought coupled with high human and cattte populations led to loss of herds and massive starvation among the people of the sub-Saharan or Sahel grasslands. Some climatologists trace this disaster to major weather trends, but drought, overgrazing, and disaster are not strangers to these or most other arid grasslands.

DESERTS

The story of the desert has one theme—water. The distribution of water in time and space determines the nature of fragile desert ecosystems. An average rainfall of less than 25 cm (10 in.)

a year will produce a desert; but each desert is different, just as are each forest and each grassland. Relative humidity, temperature extremes, the underlying rock and drainage—all contribute to the uniquenses of each desert.

Deserts are never totally dry; there is always some water. Although dew can be of considerable importance in the water budget of certain organisms, most desert life relies larger on sporadic rainfall. Particles of desert soil hold surface layers of water, as do other soils. Any excess water flows downward to collect over impenetrable layers of rock. Ground-water lying near the surface produces springs and oases. Substantial layers of groundwater underlie many deserts; some are believed to represent the accumulations of thousands or even millions of years, Perhaps this "fossil water" can be traced to rainfall during periods when these lands that are now deserts had more humid climates. Or groundwater from regions with higher rainfall (such as adjacent mountains) may feed into the rock layers under deserts. In any case, water can be drawn from desert wells faster than it is replaced.

Only rarely does vegetation carpet the desert. Consequently, rains create flash floods that dig deep gullies and carve rugged landforms in the desert. Some desert soils developed where we find them; the wind has carried others hundred of miles from their origin. Desert soils are surprisingly fertile; many support substantial vegetation after a good rain or when irrigated.

Desert Plants: Many Solutions to One Problem

Plants have evolved seemingly endless adaptations to desert conditions. Individuals divide the water supply so efficiently that they tap almost every drop. Most desert plants exhibit special modifications to gather and conserve water, but many merely become dormant during dry periods.

For example, lichen on desert rocks are active only intermittently. When dehydrated by hot desert days, these lichens suspend most life functions and for a time tolerate extremely high temperatures (to 80°C or 176°F). But after being dampened by night dews, these same plants revive and resume photosynthesis at dawn.

When the Rains Come. Although no higher plants withstand the severe dryness that lichens undergo, many live through drought as seeds, bulbs, or other low-moisture structures. Such plants usually have shallow roots that grow laterally through the top soil for considerable distance. The wide distribution of roots permits them to adsorb substantial amounts of water from a light shower.

In more moist deserts, seeds of annuals germinate every year, and the plants bloom predictably, usually within a few weeks in early spring. But in many deserts, bloom is irregular. Certain species appear only every 10 to 20 years. They survive because their seeds remain alive but inactive in the soil. Seeds of some desert plants germinate any time there is sufficient moisture, but others rely on one or more of such additional cues as light and temperature.

The short-styled sedge of the Negev in Israel responds to rain in another way. Although the old leaves lie dead on the surface of the ground, their base remain alive and start to grow right after a rain. The desert becomes green immediately with new sedge leaves, but each supports dead material at its tip.

Every Drop Counts. Root systems of perennial desert plants vary markedly. Some rely on extensive near-surface roots, especially when growing on hills or ridges far above the groundwater. Others, particularly plants near dry stream beds, produce long taproots that reach far into the soil and rock. Plants of stony deserts may send a single root under each nearby rock. This way they can absorb water that condenses from vapour during the cool desert nights.

Some cacti sprout new roots within a few hours of a heavy

rain. Once the water is absorbed and stored in the thick fleshy stem, the new roots shrivel and dry up.

Around many desert shrubs, such as the creosote bush and sagebrush, lie areas barren of vegetation. These species secrete toxins that accumulate in the soil and inhibit other species. Thus these shrubs can use all nearby moisture without competition.

Many succulent (fleshy) desert plants, such as sedums and aloes of Africa and agavas of North America, store water in thick, permanent leaves. But whether the water storage tissue is modified leaf or a stem, as in cacti, it is usually green with photosynthetic pigments. In bright, unshaded deserts, sunlight penetrates to the deepest cells. Packing the photosynthetic tissues in stems or thick leaves petmits sufficient photosynthesis and reduces the surface area from which water can evaporate.

Few desert perennials have large leaves. Some, like the gray sagebresh, grow bigger leaves in the moist season and progressively smaller ones with the coming of dry weather. The last leaves formed are mere scales. Other species, such as crown of thorns of Africa, the ocotillo of the U.S. Southwest and the boojum tree of Mexico, send out ordinary leaves during the moist season but drop them completely as soon as water is in short supply.

Special mechanisms doubtless regulate stomata in many desert plants. For example, agaves open their stomata only at night; during this time they absorb carbon dioxide and store it for use the following day. This is the reverse of the common pattern; plants of moist environments open their stomata during the day and close them at night.

Much dry-adapted vegetation bears spines or hairs. The adaptive value of these structures has been the subject of speculation. Obviously spines deter some animals from feeding on the water-rich tissues. Spines also shade and insulate the plant. When hairs are dense, they interfere with air movements near the surface. This reduces evaporation and holds an insulating layer of air that can slow transfer of heat to the plant as the environment warms. Perhaps these spines and hairs also provide a surface on which water vapour condenses into dew.

Despite all these adaptations, desert plants can suffer severe water deficits during midday temperature peaks. Such daytime deficits are made up during the night but they may actually be beneficial as physiological acclimatization to extreme heat. It is well established that enzymes are less sensitive to denaturation by heat in dry than in most environments. Perhaps midday dehydration permits desert plants to tolerate temperatures that would "cook" other vegetation.

Animal Adaptations to the Desert: A Tale of Water Conservation

Desert animals obtain their limited water rations from dew, occasionally from drinking water, but very largely from the water released during cellular respiration. In fact, this may provide the entire water supply of kangaroo rats. (You will recall that the respiration of glucose yields both carbon dioxide and water.) Birds that eat insects may also obtain sufficient water from their food. But those with a diet of dry seeds fly many miles daily to reach drinking water. Mammals, too, search for water. Some, such as the ibex and gazelle, will dig deep into the bottom of dry springs or temporary stream beds to reach groundwater.

Whatever their water source, desert animals have evolved strategies to conserve water. Many, including some reptiles and a multitude of small rodents, are nocturnal. They spend their days in cool underground burrows, where they waste little water in evaporative cooling. Although desert birds are active during the day, they seek shelter in the shade of rocks, cliffs, or plants.

The camels, donkeys, sheep, and goats that people raise in arid lands grow thick woolen coats that insulate these animals from the desert sun. Both donkeys and sheep keep their body temperatures in check through evaporation associated with painting. Consequently, they must tolerate tremendous dehydration, because they use so much water in evaporative cooling. In fact, a well-dried-out donkey can guzzle a quarter of his body weight in water within only a few minutes. Instead of using water in panting, goats and camels tolerate a rise of several degrees in their body temperatures. They accumulate heat during the day and dissipate the stored heat during the cool desert night.

Most desert rodents excrete as little water as possible with their wastes. Perhaps you have noticed the milky urine of a pet hamster. These natives of the Syrian deserts reabsorb to much water from its kidney tubules that the minerals precipitate and give its urine a milky appearance.

A large surface area-to-volume ratio could make dehydration a serious risk for desert insects that are active during the day. Many such species have evolved water-conservation adaptations that parallel those of desert plants. A thick waxy cuticle covers the bodies of many desert insects, just as desert plants have an extremely thick waxy cuticle on their leaves. All insects have multiple openings for gas exchange, not unlike the stomata of plants. But the gas exchange openings of some deserts are recessed in pits, just as are the stomara of desert plants.

CHAPARRAL

Variations in terrestrial environments seem endless. Grasslands naturally merge into deserts, and a little abuse will make deserts out of dry grasslands. Similarly, deserts grade into forests. Indeed, dry-adapted trees, such as pistachios and acacias, characterize certain deserts. Some dry, warm environments support thorn forests. The shrubby, leather-leafed plants that now cover much of the North Africa and Southern Europe are similar to the chaparral of the south-western Uunited States.

As far as can be determined, chaparral is the climax vegetation along the coast of Central and Southern California, where summers are dry but where winter can bring up to 50 cm (20 in.) of rain. In this mediterranean-type climate the chamise, manzanita, and other shrubs bear leathery leaves the year around. Severe dehydration during the dry season is largely avoided by simply closing the stomata. Nevertheless, the plants suffer water shortages during the summer. Extensive root systems penetrate as far as 8 meters (25 feed) into the soil.

Fire sweeps the chaparral regularly, destroying all vegetation above the ground. The roots of many plants survive, and new sprouts appear quickly, attesting to the availability of water in the soil. In other species, heat activities seed germination.

The chaparral community is apparently a fire climax,. Most of the minerals of this ecosystem are converted into ashes during periodic fires. Minerals leach from the ashes when it rains. This loss keeps the soil poor. Almost all chaparral plants have mycorrhizal associates that gather minerals, and many have nitrogen-fixing symbionts.

Chaparral soils are subject to heavy erosion. Homes built in chaparral areas many survive a fire, only to be washed out in a sea of mud and water when rain falls on the denuded soil.

LAND AND PEOPLE

Terrestrial ecosystems vary in their resilience, that is, in their ability to withstand abuse and to grow back when the abuse ceases. If protected from unusual erosion and loss of seed sources, most temperate forests can support successional stages and reestablish themselves in a few centuries. Grasslands seem more fragile, perhaps because it is so difficult to prevent erosion in semiarid areas. Although we have less data concerning the resilience of tropical forests and tundra, they are apparently even less stable ecosystems.

In any case, it is sobering to realize how drastically people have altered terrestrial environments. The degree of destruction in most areas seems roughly proportional to the length of time that agriculture has been practiced and to the state of industrial development. During the past 10 years, timber shortages in temperate areas have stimulated heavy logging of tropical forests. Also, well-intended but misguided attempts to increase food production have promoted clearing of tropical forest lands. Meanwhile, development of such arctic resources as petroleum is creating traffic that gouges up the tundra. All these assaults on terrestrial ecosystems trace to increasing human populations. More people require more food and more lumber and more raw materials.

In evaluating the impact of growing human numbers, we must include overall effects on the biosphere. Since the continents are a major portion of this great, worldwide ecosystem, reduced resilience in terrestrial ecosystems must contribute to a generally less stable biosphere. This reason alone is sufficient to cause concern. If people are to survive, we must take care of our precious terrestrial ecosystems.

7

Aquatic Ecosystems

About three quarters of the earth is covered by oceans, lakes, ponds, rivers, or streams. Not only is water all around us, but it is underground, too. When it rains, part of the water runs off into streams, but much of it seeps into the soil. As it does, particles absorb and hold some water near the surface. Usually it is this soil water that plants use. Some of the water trickles down through subsoil until it is eventually stopped by an impenetrable barrier of rocks. Unable to drain further, groundwater accumulates on top of the impervious rock.

The upper surface of groundwater is the water table, which more or less parallels the surface of the land. Water fills every pore and crevice between the water table and the underlying rock. Swamps, springs, and other bodies of water develop wherever the water table is high enough to intersect the surface.

Groundwater, rainwater and runoff water all participate in the water cycle driven by the sun. Solar energy evaporates tremendous amounts of water every day. This evaporation draws water from the soil, from plant leaves through transpiration, and from each body of water, large or small. Once in the atmosphere the water falls again to earth as rain or snow. Thus there is continuous recycling of moisture between earth and atmosphere and back again.

Falling rain and snow are powerful erosive forces. Most landscapes, even those of arid countries, are carved by water action. Water freezing in tiny crevices splits mighly boulders, and glaciers grind wide, smooth valleys between knife-sharp ridges. Violent streams dig narrow canyons that in time become flat plains through which a much-tamed waterway meanders to the sea. Without doubt the sun that powers the water cycle and the precipitation that condenses from water vapour are among the mightiest forces on earth. Oceans, lakes, and rivers are only a part of this great system.

THE OCEANS: THE BIGGEST PART OF THE WORLD

Water returning to the oceans from the land carries huge amounts of sediment. The buildup of these sediments, together with erosion of shores by pounding waves, has created the continental shelves that border the coasts. In general, a sharp drop, the continental slope, marks the edge of the continental shelf and the boundary of the deep oceanic basins. The shelf, slope, and in many places the ocean bottom are covered with a fine mud or ooze consisting of silt, minerals precipitated from sea water, and the microscopic shells of dead marine animals.

Much of the ocean floor is a broad abyssal plain. Here and there the floor is studded with underwater mountains, often of volcanic origin. Where these seamounts just above the water they form small, isolated islands, such as those of Hawaii. Also interrupting the flatness of the abyssal plain are deep trenches and mountainous submarine ridges. The Japanese islands are the exposed tops of one such ridge.

Drifting Continents and Changing Oceans

Recently it has been widely agreed that the continents were once jointed together, and that the present oceans were formed after the continents drifted apart. This is the theory of plate tectonics, or as it is more popularly known, continental drift. There is a significant body of evidence in support of this theory. Equally important, it explains many previously puzzling features of the earth's surface.

Close inspection of a world map reveals that some continental coastlines have complementary shapes (Fig. 7.1). Like pieces of a jigsaw puzzle, they seem to fit together. Furthermore, the present occean floors appear to be relatively young. Examination of cores drilled from bottom sediments suggests that the Atlantic Ocean formed only about 200 million years ago. That was probably when the Americas began to separate from Africa and Europe. The separation was a slow process, which still continues.

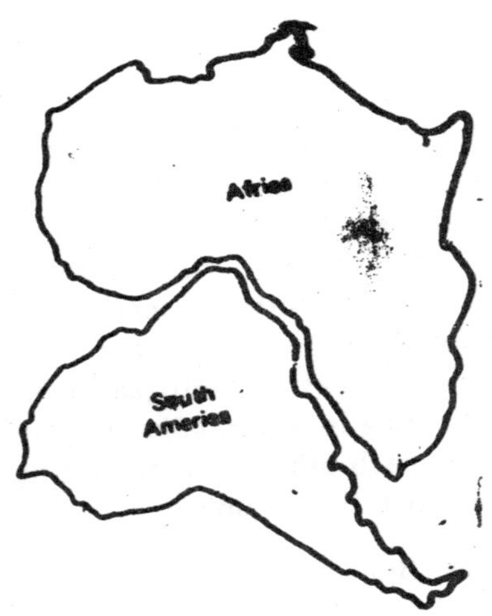

Fig. 7.1: Continental Fit. Although the physical fit of South America and Africa is quite remarkable if one considers only the surface features as shown, the result is even more striking if the undersea continental margins at the edge of the continental slopes are taken into consideration. Using this same technique, one finds that the eastern coast of the United States fits very nicely along the western shore of Africa.

Drifting of continents from one climatic zone to another may be part of the explanation for why dinosaurs and lush tropical vegetation once existed in what are now temperate areas of the United States and Europe. It certainly explains distribution of closely related species on distant continents—for example, the southern beeches found only in Chile, New Zealand, Tasmania, and Australia.

Are you asking yourself how continents move?

Crustal Plates. It is clear that the earth has an unstable crust. Catastrophic earthquakes and volcanic eruptions all too often reveal the tremendous forces at work inside the earth. Luckily for us, most internal stresses, such as those that fold the earth's crust into mountain ranges, act so slowly that they do not affect us during our lifetime. Yet these forces move continents.

Present information indicates that the earth's crust is not an unbroken skin. Instead it consists of at least six—and probably more—fairly rigid blocks or plates. These plates do not correspond exactly to any of our customary geographic boundaries but often include both continental masses and ocean bottom. Being composed of rather light rock, these crustal plates float on top of the heavier, more plastic material beneath. And like ice cubes floating in water, adjacent plates rub together and slide past or over each other as they are pushed apart by upheavals of molten rock from below.

Tectonics Means Building. The Atlantic Ocean floor is spreading as the result of activity at mid-oceanic ridges. Here new crustal material seems to be seeping out from fissures and forming additional sea floor. This, in turn, pushes the continents apart. In other oceans similar crustal plates are also growing. But since the earth is not expanding in circumference, old parts of the crust must be undergoing destruction. This seems to be happening at Pacific Ocean trenches, among other places. Such oceanic trenches, some more than six miles deep, develop when one plate is focused under another.

In places, molten crust comes to the surface, forming volcanoes. Volcanoes are particularly common along ocean ridges at places where the plates are growing. Volcanoes also appear near trenches where one crustal plate is slipping over another. In such locations, strings of volcanoes create island arcs. Not surprisingly, earthquake zones are associated with these island arcs as well as being common along other plate margins. Collisions between plates are also responsible for folded mountains. The union of India (which once was an island) with Asia is believed responsible for the crumpled crust we know as the Himalayas.

Crustal movements are an ongoing characteristic of the earth, extending far back in time. Before the Old and New Worlds were united in one supercontinent, there was a time when they were separated by a body of water called the Protoatlantic Ocean.

Considering the behaviour of the crustal plates, no marine environment is changeless, although some are exceptionally stable compared with most habitats.

Ocean Habitats

Salts are constantly added to the ocean in river water and rain-water runoff. Depending on drainage, currents, and temperature, the saltiness of the oceans varies at different locations. The average salinity is about 3.5 per cent, or about 35 parts by weight of salt to each 1000 parts of water. Most of the salt in seawater is ordinary table salt (sodium chloride), but sulphate, magnesium, calcium, and potassium salts are also abundant. In fact, seawater contains traces of just about every element.

These accumulated salts make seawater much denser than fresh water. This is the reason it is easier to swim or float in the ocean than in a river or a lake. It also means that salt water has a much lower freezing point than fresh water. In addition to salt, seawater also contains dissolved atmospheric gases. But seawater has less oxygen than fresh water, since salt decreases the solubility of oxygen in water.

Ocean Layers and Life. The sun heats and illuminates the ocean, although few of the sun's rays penetrate very deep into the water. As a result, water temperature, illumination, and salinity (which is temperature dependent) are related to depth. Of course, all these characteristics influence the kind of organisms to be found at a given location, but light is of special importance.

Because photosynthetic plants need light in order to live, they are found only in shallow water and in the surface layer of the open ocean. Almost all photosynthesis occurs in the upper 80 meters (260 feet) designated as the euphotic (good light) zone. Underneath the euphotic zone is a region of ever-deepening twilight. No sunlight penetrates below 600 meters (1967 feet). As a result, most of the ocean is in perpetual darkness and can be inhabited only by animals and decomposer bacteria and fungi.

Life at the Top. Microscopic floating algae, known collectively as phytoplankton, are the main producers in the oceans. Diatoms and dinoflagellates are the most prominent phytoplankton. Their collective biomass far outweighs that of the more obvious algae known as "seaweeds." Most seaweeds are attached to rocks near the coast.

Phytoplankton are never evenly distributed. Not only are they limited to the top layers where there is light, but within that layer their density follows distribution of minerals. Organisms that die in deep water usually drift to the bottom. Similarly, faces settle out, and the valuable nutrients they contain are lost from surface waters. Currents and diffusion only slowly return precious nitrogen and phosphorus from the bottom to the euphotic zone. Shallow water over the continental shelves is usually fairly well mixed and contains more nutrients than surface water over the open ocean. Here the deep water is quite undisturbed, and minerals are concentrated far below the euphotic zone.

During the winter, because sunlight striking temperate waters is less intense than at other seasons, photosynthesis is limited to the top few meters of water. With the coming of spring the phytoplankton multiply rapidly in response to increasing illumination. Exploding plankton populations are termed plankton blooms, because the masses of algae often colour the water bright red, brown, or green. Although clearly visible, marine plankton blooms are never as dramatic as those of freshwater lakes. This difference in productivity may result from mineral deficiencies in illuminated marine waters.

Microscopic animals, often referred to as zooplankton, feed on phytoplankton. Copepods and krill, tiny relatives of crabs and shrimp, are among the most abundant zooplankton. Others are numbered among the unicellular protozoa and include foraminifera and radiolarians. Typically, zooplankton bear numerous projections that increase their surface area and help suspend them in the water. Most zooplankton both swim and float.

For the most part, relatively small fish, such as anchovies, herrings, and sardines, prey on zooplankton. There are exceptions, however. In the cold but fertile water around Antarctica the penguins, fish, squid, and even baleen whales feed directly on krill. The amount of zooplankton consumed by a large animal can be astounding. A single blue whale can eat three tons of krill a day.

The Briny Deep. Beneath the productive upper layer of the ocean lies as much as six miles of water. Throughout most of this space, life is sparse. Food from the top decomposes quite slowly.

Fish in deep, dark waters are mostly mouth. Good meals are rare here and few items are rejected as being too big. This is the province of the famous angler fish, which is equipped with iuminescent lures that attract prey into its gaping mouth. The low population levels in deep water makes finding a mate at the right time a chancy proposition. In some species the males attach to the females during adolescence and maintain this parasitic relationship throughout life. As you might expect, the males of such species are tiny compared with their mates.

Rocks and Sand: Marsh and Muck

In contrast to the relatively uniform conditions of the open sea, a variety of habitats occur where water meets land. Each of these environments is unique; each is intriguing in its own way. Perhaps rocky shores are the most interesting. Here we find a diverse community attuned to the rhythmically changing environment.

Living between the Tides. In the band that spans high and low tide, the first priorities are to stay put, to keep wet, and to avoid being crushed. Waves pound the shores so vigorously that delicate organisms can be destroyed or carried away. Alternately submerged and exposed, intertidal species (those that live between lowest low tide and highest high tide) must also withstand cold, heat, and a wide range of salinity. Summer sun can cook tissues, and evaporation of seawater deposits a crust of salt on every surface. Only a few months later, winter tides pull the protective layer of water away and leave organisms exposed to freezing cold or to torrents of fresh water. As adaptations to this extreme environment, intertidal organisms either cling tenaciously or dart about to find shelter. Their bodies are protected by shells, tough body walls, leathery surfaces; or muscus.

Since tidal exposure varies every day, there are distinct life zones along the shore. The following applies to the central California coast, but similar areas occur on rocky shores from Mexico to Alaska and along the coast of Maine and the Maritime Provinces, as well as on other continents.

A dark band of cyanobacteria (blue-green algae) marks the highest zone where there is much marine life. Here numerous periwinkles scavenge. A distinict band of white barnacles grows below the black zone. Barnacles lie on their backs, stuck to the rocks. When the tide goes out, barnacles close their limy shells to conserve water. When they are submerged, they open their shell plates and feed, using delicate, jointed appendages to kick food into their mouths.

Limpets are abundantly distributed among the barnacles. These marine snails use a tremendous suction foot to pull their shells tightly against the rock. This traps water under the shells and prevents dehydration during low tide.

A crowded strip of mussels stands just below the mean-sea-level mark that coincides with the bottom of the barnacle region. The clamlike mussels spin proteinaceous fibers that attach their blue-black shells to the rocks. Tidal pools harbour rock crabs, hermit crabs, and large green sea anemones. Farther out, on the underside of large rocks and steep ledges hide sponges, sea cucumbers, sea squirts, chitons, and abalone. Shrimp and spiny lobsters lurk in the lowest tide pools. No one has made a complete census of animals in the intertidal zone, but bioiogists familiar with the California coast estimate that there may be 3000 species there.

The showy life of rocky shores appears missing from sandy beaches, but even though beaches appear barren, animals are present. It's just that most of them are hidden from view. Burrowing, digging, and tunnelling creatures, among them clams, mole crabs, tube worms, and olive shells, bury themselves in the sand. Low tides expose some species periodically; others are permanently protected within their burrows. Typical residents below the low-tide line include whelks, swimming crabs, sand dollars, and hermit crabs.

Wetlands and Estuaries. Bays and river mouths, where fresh and salt water meet, are unusually fertile environments. The most valuable of such estuaries are broad, shallow basins created by silt deposits. Here mud flats and tidal marshes line channels of open water. Most of the silt and organic matter dumped from the river becomes trapped in its estuary. Slow currents, a large surface for evaporation, and presence of rooted vegetation all help make estuaries places where nutrients tarry.

Salt marsh grasses, eel grasses, and other rooted plants take advantage of rich water and soft bottoms. Algal scums cover mud flats, larger plants, and every solid surface, and dense phytoplankton blooms colour the water. There is so much food that only a tiny part of the photosynthetic product finds its way directly into the mouths of herbivoures. Instead, most plants die and, in the absence of swift currents, simply sink to the bottom and rot. The abundant organic matter supports a teaming broth of bacteria and fungi that use all available oxygen from the muddy bottom. As a result, other anaerobic micro-organisms thrive there and release hydrogen sulphide. This noxious gas blackens marshland mucks and imparts a characteristic stench.

Their tremendous photosynthetic productivity and extensive decomposer food chain permit estuaries to literally teem with life. Microscopic decomposers are eaten by filter-feeding zoo-plankton, as well as by larger filter feeders, such as worms, clams, and oysters. Estuaries serve as nurseries for numerous coastal fish. Menhaden, striped mullet, summer flounder, king whiting, croakers, striped, bass, smelt, and sturgeon are all commercially important species that rely on estuaries sometime during their lives. The shrimp fisheries also depend on this resource, because young shrimp require the estuary habitat. Oysters, blue crabs, and several commercially valucable elam species are permanent estuarian residents. Furthermore, salt-water marshes are the natural home of most waterfowl.

As valuable as they are in their native state, estuaries are often prized more for their land. To take advantage of a protected harbor, most seaports are built on estuaries. As early ports grew, the surrounding marshes and mud flats were usually drained, and regions of shallow water were dredged or filled to provide areas for docks, industries, airports, and even residences. In Washington, D.C., only the Tidal Basin remains to remind us that the Jefferson Memorial, Lincoln Memorial, and

Washington Monument were built on land "reclaimed" from a marsh.

Once leached free of their load of salt, drained tidal marshes become fertile farmland. Over the centuries, carefully engineered diking and pumping schemes created much of the Netherlands. But present generations seem little aware of how much humans are responsible for familiar shorelines and of how important estuaries are in the schemes of aquatic life. Even now, many of our least-modified estuaries are threatened with "development" into resorts.

LAKES: QUIET CHANGES

In many ways lakes resemble oceans. Of course, lakes are much smaller, and they usually contain fresh water. Diatoms and desmids are the most abundant phytoplankton of lakes. And though the shoreline and broad-leaved floating plants contribute much more to the productivity of fresh waters than marine seaweeds do to the oceans, zooplankton and fish are abundant in both habitats. Probably the most distinctive animal components of the freshwater food chain are the many insects, especially in their juvenile stages.

Variations in climate, fertility, shape, and bottom characteristics cause lakes to be quite different from one another. But ultimately all these factors translate into nutrient availability. Unlike the oceans, temperature zone lakes are stirred regularly. As a result. there is efficient nutrient cycling. This is what accounts for the much greater productivity of many lakes compared with any ocean.

Seasonal Turnover

Swimmers know that lake water is warmer at the surface and colder at depths. Thus in the summer, lakes are stratified with lighter, warmer water at the top and colder, denser water toward the bottom.

As winter approaches, falling air temperatures and brisk

winds slowly cool temperature zone lakes. The fact that fresh water is most dense at 4°C (39°F) permits cooling lakes to be thoroughly stirred. When surface waters cool to 4°C, they sink to the bottom and displace the warmer, less dense water there. This turnover brings nutrients from the bottom and spreads them throughout the lake. At some point the entire lake is at 4°C and lacks any density stratification. Absence of stratification allows the lake to be further stirred by the wind. Only after the whole lake reaches 4°C may the temperature of the surface dip lower. When this occurs, the colder, less dense water floats and a layer of ice may eventually form.

In the spring, the surface water gradually warms to 4°C. Then, as during the fall-cooling, there is a period when all the water is 4°C and all of the same density. At this time winds blowing across the surface may set up currents that travel toward the shore and turn downward and move across the bottom.

As you can see, there are two turnovers a year, one in the spring and the other in the fall. Both stir bottom sediments, and both also replenish the oxygen supply at the bottom. With available nutrients and increasing sunlight and warmth, plankton blooms colour lake water beginning in the spring. Changes in temperature and nutrient levels cause one phytoplankton species to bloom and die, only to be followed by a population burst of another. The exact succession of blooms is determined by the requirements of the various species.

Nutrients, Pollution, and Aging

Lakes have a relatively short life span. It may take thousands or millions of years, but most lakes that we know are in the process of disappearing. At the time of formation, a young lake is usually low in nutrients and sparsely inhabited. As time goes by, the lake is enriched by accumulation of river sediments and nutrients leached by rainwater from the surrounding land. Further fertilization occurs as dead plants fall into the water and decay. This normal enrichment of a lake is known as eutrophication (addition of good food). Eutrophication permits more and more organisms of increasingly different kinds to live in the lake. Clearly, the biological nature of a lake gradually changes as the lake ages. With time, lakes undergo a process analogous to succession on land. Eventually eutrophication may end with disappearance of the lake altogether. The Fate of Lakes. As sediments build up, lakes become more shallow and vegetation encroaches. Filled-in lake margins develop into marshes or swamps. By definition a marsh is a wet grassland and swamps are wet forests. Like their saltwater counterparts, freshwater marshes support diverse communities adapted to slight variations in water levels. So do swamps. And like marine wetlands, those of fresh water are gradually disappearing. Conversion of the rich muck into farmland is the goal of many government-funded drainage and stream channelization projects.

In cold climates lakes may eventually turn into spongy, wet bogs. Because of the cold, accumulated plant material decomposes poorly, and it is gradually transformed into peat. Humic acids derived from the peat further inhibit decomposition and give the water a characteristic brown tinge. Peat is often dug up, dried, and burned as fuel.

Excess Eutrophication. Although natural eutrophication is part of the aging process of every lake and tends to favour mineral accumulation, human activities that accelerate the process can lead to early death of a lake. Fertilizer washed into lakes from nearby farms and the addition of minerals in household or industrial sewage can create an overabundance of nutrients in the water. Under these conditions algal blooms deplete oxygen in the water. Fish and other animals may simply choke to death.

Studies of the Great Lakes trace their "eutrophication" to excess phosphorus. Experimental addition of phosphate to small lakes in the same region has produced responses similar to pollution in the Great Lakes. Neither nitrogen nor carbon alone has a similar effect. Apparently simply removing phosphate from waste water, principally by banning phosphates in household detergents, would markedly improve conditions in many bodies of water.

RIVERS AND STREAMS: RUNNING WATERS

The continuous one-way flow of running waters distinguishes them from lakes. Indeed, many of the special characteristics of these waters are related to their turbulence. Because currents constantly bring new water into contact with a given organism, running water is effectively richer than still water in which a static, "tired" layer surrounds each individual. And to be sure, turbulence mixes air into running water and thus usually ensures a plentiful supply of oxygen.

The Food Web of an Open System

Rivers and streams inevitably reflect the land they drain. Stream water arises mainly from groundwater, and surface runoff provides most of the remainder. Water from both sources carries mineral nutrients and human additives, such as pesticides. But most of the organic matter that supports the food chains of running waters also washes in from the surrounding land. Or simply falls in. Probably tree leaves supply more energy and more carbon than any other single source. Additional sources of carbon and energy ase adult insects that flounder into the water, earthworms, and other small soil animals. Except in quiet streams that closely resemble lakes or ponds, there are few phytoplankton in rivers. Thus running waters are unusual ecosystems in that primary production accounts for but a small portion of the energy needed. Most nutrients enter rivers and streams from the outside.

Adaptations of Currents: Dont's Get Carried Away!

Only careful strategies prevent the current from carrying stream organisms to different and less suitable habitats. For example, diatoms of such habitats have stalks. Larvae of Blackflies and many other insects anchor themselves with holdfasts. Yet other insects spin sticky threads about themselves or burrow into the mud. Many stream animals are streamlined. They have compact and flattened bodies and usually live under stones, where the current is reduced and where they are less visible to predators.

Fish can expend a great deal of energy swimming against the current—so they don't. Fish survive by finding shelter. As anyone who fishes knows, the best place to catch trout is in a deep bottom hole or along the downstream side of large rocks or other obstacles. Generally, the larger the shelter, the larger the fish.

The adaptations of stream organisms truly illustrate the selective forces of this environment. But of course, every environment is selective.

A SPECIAL WAY OF LIFE IN WATER

To an animal, the most fundamental difference between living on land and living in water is the scarcity of oxygen. Oxygen is so poorly soluble in water that it is hardly surprising that no warm-blooded animals can live submerged for more than short periods. Aquatic mammals, such as whales, and aquatic birds, such as penguins, all breathe air and thus obtain enough oxygen to sustain the high metabolic rate associated with a constantly warm body. And surely the complexity of gills is related to the need for extensive surfaces for gas exchange with water. Most often gills are located where water can be readily pumped past them to maintain a constant supply of oxygen. As we will see in the next chapter, the deadly effects of aquatic pollution can be traced to exhaustion of the oxygen supply.

Aquatic ecosystems are also unusual in that most producers are microscopic Since so many people disregard anything they can't see, this may be why the structure of underwater communities escapes the understanding of many citizens. Microscopic producers dominate in water, because they have a large surface area compared with their volume. This ensures efficient absorption of nutrients. Furthermore, vascular tissue is unnecessary in this habitat, and large plants are subject to damage by waves or current.

Finally, the high specific heat of water gives aquatic environments a temperature stability unknown on land. Water is slow to warm and slow to cool. Large bodies of water, especially, fluctuate little in temperature. Not only are aquatic organisms freed from the usual problems of temperature regulation, but density changes associated with minor temperature shift account for lake turnover and also affect ocean currents. Remembering these facts can help us maintain the integrity of aquatic ecosystems. This is an important step toward ensuring stability of the biosphere.

8

Light and Plant Development

Various kinds of plant germinate, grow, flower and fruit at different times in the year, each in its own season. Thus some plants flower in the spring, others in the summer and still others in the autumn. And in the autumn, trees and shrubs stop growing in apparent anticipation of winter, usually well before the weather turns cold. What is the nature of the clock or calendar that regulates these cycles in the diverse life histories of plants? Some 40 years ago it was discovered that the regulator is the seasonal variation in the length of the day and night. Since this is the one factor in the environment that changes at a constant rate with the change of the seasons, in retrospect the discovery does not seem so surprising. But how do plants detect the change in the ratio of daylight to darkness? The answer to this question is just now becoming clear. It appears that a single light-sensitive pigment, common to all plants, triggers one or another of the crises in plant growth, from the sprouting of the seed to the onset of dormancy, depending upon the plant species. This discovery is a major break-through toward a more complete understanding of the life processes of plants, and it places within reach a means for the artificial regulation of these processes.

The pigment has been called phytochrome by the investigators who discovered it at the Plant Industry Station of the U.S. Department of Agriculture in Beltsville, Md. It has been partially isolated, and it has been made to perform in the test tube what seems to be its critical photosensitive reaction: changing back and forth from one of its two forms to the other upon exposure to one or the other of two wavelengths of light that differ by 75 millimicrons. A millimicron is a ten thousandth of a centimeter). Phytochrome appears to be chemically active in one of its forms and inactive in the other. In the tissues of the plant it functions as an enzyme and probably catalyzes a biochemical reaction that is crucial to many metabolic processes.

The first step toward the discovery of phytochrome was taken in the 1920's, when W.W. Garner and H.A. Allard of the Department of Agriculture recognized "photoperiodism" [see "The Control of Flowering," by Aubrey W. Naylor; Scientific American Offprint 113]. They showed many plants will not flower unless the days are of the right length—some species flowering when the days are short, some when the days are long. Indeed, some plants seem to react not simply to seasonal changes but to changes in the length of the day from one week to the next. Photoperiodism explained why plants of one type, even though planted at different times, always flower together, and why some plants do not fruit or flower in certain latitudes.

Other investigators soon reasoned that if a plant needs a certain length of day to flower; then keeping it in the dark for part of the day would inhibit its flowering. They tried to demonstrate such an effect in the laboratory. Nothing happened: the plants always bloomed in the proper season. The riddle was solved when the reverse experiment was tried, that is when the night was interrupted with a brief interval of light. Chrysanthemums, poinsettias, soybeans; cockleburs and other plants that flower during the short days and long nights of autumn and early winter remained vegetative (*i.e.*, non-flowering). Moreover, they could be made to bloom out of season, when the night was lengthened by keeping them in the dark at the beginning or the end of a long summer day.

Conversely, a brief interval of light interrupting the long winter night induced flowering in petunias, barley, spinach and other plants that normally bloom in the short-night summer season. Artificial lengthening of the short summer night kept these plants vegetative. Interrupting or prolonging the night-time darkness correspondingly affected stem growth and other processes as well as flowering in many plants.

It was evident that light must act upon a photoreceptive compound, or compounds, to set some mechanism that runs to completion in darkness. As a first step toward elucidating the chemistry of the process H.A. Borthwick, Marion W. Parker and Sterling B. Hendricks of the Department of Agriculture in 1944 set out to determine the wavelength or colour of light that is most effective in inhibiting flowering in long-night plants. They exposed each of a series of Biloxisoybean plants, from which they had stripped all but one leaf, to different wave-lengths of light from a large spectrograph [as in middle illustration on next page]. Several days after the treatment the plants were examined for the effect of this exposure upon the formation of buds. Red light with a wavelength of 660 millimicrons proved to be by far the most effective inhibitor of flowering. The cocklebur and other long-night plants gave the same response. By plotting on a graph the energy of light required to inhibit flowering at various wavelengths, the investigators obtain-the "action spectrum" of the mechanism that inhibits flowering. This showed that to interfere with flowering, much more light energy is required at, for example, 520 or 700 millimicrons than at (or very near) 660 millimicrons. The wavelength at which the unknown substance absorbs light most efficiently was thus shown to be 660 millimicrons.

Borthwick and his associates then turned to short-night plants. The same spectrographic experiments yielded exactly the same action spectrum. But in this case the effect of the exposure was to promote—not inhibit—flowering! Since the same wavelength of light caused the greatest response in both cases, the investigators could only conclude that a single photoreceptive substance was involved in these two diametrically opposed responses.

Subsequent experiments implicated the same compound in the control of stem elongation and leaf growth. Recent work has shown that light at 660 millimicrons also acts upon mature apples to turn them red by enabling them to make the pigment anthodyanin. The same red light controls the production of anthocyanin in a number of seedlig plants.

With the collaboration of a research group headed by Eben H. Toole, Borthwick and his associates next set out to determine which wavelengths of light trigger germination in those seeds that must be exposed to light in order to grow. Many weed and crop seeds are of this type. The action spectrum for the promotion of seed germination turned out to be essentially the same as that established for other plant responses. Whereas about 20 per cent of the seeds germinated in the dark or when they were exposed to green, blue and other shorter-wavelength colours, more than 90 per cent sprouted after irradiation by red light at 660 millimicrons.

This finding led the two groups of investigators to the study of an entirely different effect of light upon germination. It had been observed in the late 1930's that germination was inhibited when seeds were exposed to the longer wavelengths of far-red light which are invisible to the human eye. That observation was speedily confirmed, In fact, seeds that had been pushed to maximum germinative capacity by exposure to red light failed to germinate when they were subsequently irradiated with far-red light. The plotting of the action spectrum for this effect showed that far-red light at 735 millimicrons wavelength is the most potent in inhibiting the germination of seeds.

Still more interesting was the discovery that the diametrically opposed effects of red and far-red light upon germination are fully reversible. After a series of alternate exposures to light of 660 and 735 millimicrons, the seeds responded to the light by which they were last irradiated. If it was red, they germinated; if far-red, they remained dormant. Now all phenomena of growth a ıd flowering had to be re-examined for the effect of far-red as well as of red light. In each case the experiments demonstrated that irradiation by far-red light reversed the effects obtained by irradiation with red light.

The reversibility of these reactions and the clear definition of their action spectra strongly suggested that a single lignt-sensitive substance is at work in every case and that this substance exists in two forms. One form, which was designated phytochrome 660, or P_{660}, absorbs red light in the region of 660 millimicrons. When P_{660} is irradiated at this wavelength, it is transformed into phytochrome 735 (P_{735}). which absorbs far-red light at 735 millimicrons. When P_{735} is irradiated with light at 735 millimicrons, it reverts in turn to the P_{660} form.

In order to substantiate these deductions phytochrome had to be extracted from the plant. This required a method for detecting the presence of the compound other than the responses occur at sharply defined wavelengths, there was reason to expect that phytochrome itself would prove to be more opaque, or "dense," to light at the wavelengths of 660 and 735 millimicrons when examined in a spectrophotometer—an instrument that measures the intensity of transmitted light at discrete wavelengths. The transformation of phytochrome from one form to the other would also show up well.

Measuring the very small amounts of phytochrome present in plants was not a simple task. K.H. Norris and one of the authors (Butler), respectively an engineer and a biophysicist in the Department of Agriculture, had been studying the pigment composition of intact plant-tissue, and had developed some sensitive spectrophotometers that measured the absorption of light by leaves and other plant parts. With Hendricks, a chemist and H.W. Siegelman, a plant physiologist who had been investigating the chemistry of phytochrome, they formed a research group to look for the reversible pigment. Initially the absorption of light by plant parts failed to reveal the presence of phytochrome. The plant tissue, however, contained large amounts of chlorophyll, which absorbs strongly at 675 millimicrons. Apparently the absorption of light by chlorophyll—at a wavelength so near the phytochrome absorption peak of 660 millimicrons—was masking the absorption by phytochrome.

It was no great problem, however, to get around this obstacle. Seedlings can be sprouted in the dark and can grow for a while on the food energy stored in the seed, they do not begin to synthesize chlorophyll until they are exposed to the light. Corn seedlings were accordingly grown for several days in complete darkness. They were then chopped up and exposed to red light to put the phytochrome into the P_{735} form in which it absorbs far-red light. In the spectrophotometer a weak beam of light at 735 millimicrons was projected through the sample, weak light being used so that the phytochrome would not change form. The absorption spectrum showed that the phytochrome was indeed absorbing light at 735 millimicrons; the same sample passed light at 660 millimicrons. On the other hand, after exposure to relatively bright far-red light the chopped seedlings were found to absorb more light at 660 millimicrons and less light at 735 millimicrons. Repeated demonstration of this reversibility fully confirmed all that had been predicted from the responses of whole, growing plants. No doubt remained that a single compound was resposible for the reversible changes in growing plants.

The spectrophotometer measures the changes in "optical density" with such high sensitivity that it can be used to assay the amount of phytochrome in plant tissue. Thus with the help of this instrument a search was instituted for a plant that would supply phytochrome in sufficient abundance for chemical separation. Certain plant tissues, such as the flesh and seed of the avocado and the head of the cauliflower, showed a relatively high concentration of phytochrome. The cotyledons (the first leaves, which feed the seedling) of most legumes synthesize phytochrome, and the concentration reaches its maximum about five days after the seeds start soaking up water. The growing shoot, or hypocotyl, as well as the first leaves of the legumes, contain less phytochrome than the cotyledons. In cabbage and turnip seedlings that have been sprouted in the dark the cotyledons are also a good source of phytochrome. However, five-day-old, dark-grown corn seedlings proved to be the best source because they develop a high phytochrome content, have large stems and are easy to grow and harvest.

The preliminary and partial chemical isolation of phytochrome was easily accomplished. Corn shoots were ground up in a blender along with water and a mild alkaline buffer, and a clear solution was separated from the solid material by filtration. This extract exhibits exactly the same reversible optical-density changes at 660 and 735 millimicrons as the chopped seedlings themselves. Thorough study of the partially purified material has developed no evidence that any compound other than phytochrome participate in the photoreaction.

Though phytochrome has not yet been isolated in pure form, the outlook is favourable. The compound shows all the properties of a relatively stable soluble protein, and Hendricks and Siegelman are now using the techniques of protein chemistry to purify it. They have subjected it to dialysis (diffusion through a porous membrane), and it has retained its photochemical activity. Oxidizing and reducing agents do not affect it. Moreover, the photoconversion occurs at zero degrees centigrade as readily as at 35 degrees, as would be expected of a strictly photochemical reaction. Higher temperatures denature the protein and destroy its photochemical activity.

Measurements of the amount of red and far-red light necessary to bring about the conversion show that P_{660} consumes only one third as much energy in being transformed to P_{735} as P_{735} consumes in being changed into P_{660}. In both cases, however, the energy consumption is relatively small, indicating that the phytochrome absorbs light efficiently. The pigment should turn out to be a blue or blue-green, the colors complementary to red, but the concentration achieved so far has been too low to make the colour visible.

Experiments with growing plants had indicated that P_{735} slowly changes back into P_{660} in darkness, whereas red-absorbing P_{660} is stable. Direct measurement of the changes in the form of phytochrome in intact corn seedlings have confirmed these indications. In seedings that have never been exposed to light, phytochrome occurs entirely in the red-absorbing, or P_{660}, form. These seedlings are exposed briefly to red light to convert the P_{660} to P_{735}. They are then returned to the darkroom and are examined with the spectrophotometer at intervals thereafter. Such measurements show that it takes about four hours at room temperature for the P_{735} to change back into P_{660}.

This conversion in the absence of light is apparently mediated by enzymes. It is markedly retarded by lowering the temperature, and it does not occur at all in the absence of oxygen. In the partially purified clear liquid extracts of seedlings, however, P_{735} is stable in the dark, indicating that the dark-conversion enzyme system has been removed. Half the phytochrome activity is lost in intact seedings during the dark-conversion of P_{735} to P_{660}. After a second illumination with red light, total phytochrome activity declines still more. In continuous light, phytochrome activity is quite low, but is still detectable. It was fortunate that the presence of chlorophyll made it necessary to grow seedlings in darkness for the early experiments. If the seedlings had received even small amounts of light, the unstable P_{735} would have formed and would have soon lost its activity to such an extent that phytochrome might never have been detected.

How phytochrome exerts its manifold influences on plant growth is still unknown. P_{735} seems to be the active form, while P_{660} appears to be a quiescent form in which the plant can store the potentially active compound. At the close of any period of exposure to light, phytochrome is predominantly in the far-red-absorbing form. The rate at which P_{735} is then carried through the dark conversion back to P_{660} provides the plant with a "clock" for measuring the duration of the dark period. The rate of conversion probably varies from one plant to another, and must depend in part upon such factors as temperature. The effective dark period might be the time required for the complete conversion, or it might be the time in darkness after the conversion is finished. In either case a brief interval of light in the middle of the dark period would cause a plant to respond as though the dark period were short.

Phytochrome is undoubtedly an enzyme—a biological catalyst. Its ability to control so many kinds of plant response in so many different tissues suggests that it catalyzes a critical reaction that is common to many metabolic path-ways. Several reactions of this kind are known. One is the reaction that forms the so-called acetyl coenzyme-A compounds. These compounds are essential intermediates in fat utilization and fat synthesis, in cellular respiration and in the synthesis of anthocyanin and sterol compounds. The regulation to the supply of acetyl coenzyme-A compounds would provide an ideal control for growth processes. More than three fourths of all the carbon in a plant is incorporated in this coenzyme at some stage or other.

The extraction and partial purification of phytochrome is the starting point of a major forward movement in the understanding of plant physiology. Further research on this remarkable protein, ubiquitous in the plant world, should answer many questions concerning germination, growth, flowering, dormancy and colouring. It should also provide a means to control all of these plant processes to the great benefit of agriculture.

The progress of each group's flower-bud development was directly proportional to the excess of its dark period over eight and a half hours, up to a certain limit.

What does this mean? The simplest and most plausible explanation is that the plant synthesizes the flowering hormone in the late stages of the dark period, and the more time it has for synthesizing the hormone, the faster the flower bud will develop. However, diminishing returns set in after about 12 hours of darkness; in fact, it appears that when the dark period is prolonged to 20 hours or so, the hormone begins to be destroyed.

This brings us to the somewhat controversial next step. The cocklebur will often flower even if it is left in the dark for nine days. But James Lockhart and Hamner, working at the University of California at Los Angeles, have found that for full flowering development the plant seems to need a re-exposure to high-intensity light, *e.g.*, sunlight, after the dark period. Consequently it may be that this exposure represents an other chemical step in the normal flowering process: the light may be needed to "stabilize" the flowering hormone or to stop some process in the leaves that destroys the hormone.

A very simple experiment has disclosed another distinct stage in the requirements for flowering: namely, transport of the flowering hormone from the leaves to the growing stem tips where the flowers form. If a cocklebur's leaves are removed at the end of the critical dark period, the plant will not flower. This must mean that the flowering hormone is still in the leaves and none has yet reached the stem tips.

Finally, there are obviously two more steps in the flowering process; the flowering hormone acts upon the cells in the growing tips in some way to transform, or differentiate, them from the vegetative to the flowering form of growth, and the bud then develops into the specific structures that make up a flower.

To sum up, the flowering of a plant has been resolved so far into about eight steps:

1. the build-up of needed substances in the plant by photosynthesis,
2. conversion of a pigment in the leaves to the non-inhibitory form in darkness,

3. another preparatory reaction in the darkness,
4. synthesis of the flowering hormone, also in darkness,
5. a possible further chemical reaction requiring exposure to intense light,
6. transportation of the flowering hormone from the leaves to the growing stem tips,
7. alteration of the vegetative cells there to the flowering mode of growth,
8. development of the flower bud.

All this gives us only a tantalizing general view of the process, as if we were watching the building of a house and could see the foundation, floors, walls and roof go up but were too far away to see any of the details (wiring, plumbing, etc.) that make it a living home. We long for a more intimate view, which means a closer and more direct look at the chemical reactions involved in generating a flower.

At Colorado State University we have recently been doing some experiments which are opening some enticing new avenues of approach. Mainly these experiments take the form of testing the effects of various chemicals applied to the plant before, during or after the dark period.

Some chemicals, known to be general inhibitors of plant growth (*e.g.*, maleic hydrazide), have proved to inhibit flowering in our cockleburs regardless of when they were applied. Others seem to exert a more specific action. For example, the growth hormones (auxins) interfere with flowering only if they are applied before the flowering hormone has traveled from the leaves to the growing tips. Our experiments suggest that they may play some part in destruction of the hormone in the leaves. Another chemical, the growth inhibitor dinitrophenol, prevents flowering only if it is applied to the cocklebur during the dark period. We think that it probably blocks synthesis of the flowering hormone. Dinitrophenol is known to interfere with the production of energy-rich phosphate bonds. If such bonds are the source of energy for the synthesis of the flowering hormone, dinitrophenol may exert its inhibiting effect by cutting off the power supply, so to speak.

One of our most useful tools for dissecting the flowering process has been the ion of cobalt. This chemical inhibits flowering only if applied during the early hours of the dark period. Thus we can narrow down its action to interference with one of two processes: the conversion of the pigment involved in flowering or the "preparatory reaction." Furthermore, among all the chemicals tested so far the cobalt ion is the only one that affects the length of the dark period needed for flowering. In some experiments plants treated with cobaltous chloride did not flower unless their dark period was at least 11 hours long. In other words, cobalt slows down the clock that ticks off the critical night for the cocklebur.

With refined probing tools such as cobalt and other specific growth regulators we have hopes of eventually being able to pinpoint the substances and reactions that bring the cocklebur to flower. But already the experiments have opeded wider horizons. The specific wavelengths of orange-red and far-red light that so powerfully influence the flowering of cockleburs seem to control many other plant processes-not only the flowering of many plants but also the colouring of tomato skins, the germination of lettuce seeds, the growth of seedlings in the dark, the winter dormancy of tree buds, and so on. Perhaps we are on the track of a key pigment which plays a large role in the plant kingdom, and possibly even among some species of the animal kingdom.

9

The Path of Carbon in Photosynthesis

The processes of life consist ultimately of the synthesis and breakdown of carbon compounds. Because a carbon atom can bind four other atoms to itself at a time and is thereby able to link up with other atoms—especially other carbon atoms—in chains and rings, carbon lends itself to the construction of a virtually endless variety of molecules. These molecules derive their physical characteristics and chemical activity not only from their composition but also from their size and intricacy of structure. The rich variety of life suggests in turn that living cells have gone far in the elaboration of such compounds and the processes that make and unmake them. All these processes depend in the end on a first one. This is the process of photosynthesis, which takes carbon and several other common elements from the environment and builds them into the substances of life.

The plant finds most of these elements already bonded to oxygen in oxides such as carbon dioxide (CO_2), water (H_2O), nitrate (NO_3^-) and sulphate (SO_4^-). Before the plant can bind the elements other than oxygen together as organic compounds, it must remove some of the excesss oxygen as oxygen gas (O_2). and this accomplishment takes a large amount of energy.

In the simplest terms photosynthesis is the process by which green plants trap the energy of sunlight by using that energy to break strong bonds between oxygen and other elements, while forming weaker bonds between the other elements and forcing oxygen atoms to pair as oxygen gas. For example, to make the sugar glucose ($C_6H_{12}G_6$) the plant must split out six molecules of oxygen in order to combine the carbon and half the oxygen of six carbon dioxide molecules with the hydrogen of six water molecules.

The glucose and other organic compounds taken up in the chemical machinery of the plants and the animals that on plants serve both as fuel and as the raw materials for the synthesis of higher organic compounds. That considerable solar energy is bound by photosynthesis becomes apparent when wood or coal is burned. In living cells the controlled combustion of respiration extracts this energy to power the other processes of life. Both kinds of combustion take oxygen from the air and break down organic compounds to carbon dioxide and water again. In its end result photosynthesis can be defined as the opposite of respiration. Together these complementary processes drive the cyclic flow of matter and the noncyclic flow of energy through the living world.

From such generalizations about the effect and function of photosynthesis in nature it is a long step to the explanation of how photosynthesis works. Yet much of the explanation is now complete. The work has been greatly facilitated by the earlier and more nearly complete resolution of the chemistry of respiration. The two processes, it turns out, are in some ways complementary on the molecular scale, just as they are on the grand scale in the biosphere. Each involves some 20 to 30 discrete reactions and as many intermediate compounds; half a dozen of these reactions and their intermediates are common to both photosynthesis and respiration. Only the first few steps in photosynthesis are driven directly by light. The energy of light is trapped in the bonds of a few specific compounds. These energy carries deliver the energy in discrete units to the steps of synthesis that follow. The same or closely similar carriers perform corresponding operations in respiration, picking up energy from the stepwise dismemberment of the fuel molecule and delivering it to the energy-consuming processes of the cell. Although the first, energy-trapping stage in photosynthesis remains to be clarified, it is now possible to trace the path of carbon from the very first step in which a single atom of carbon is captured in the bonds of an evanescent intermediate compound.

The term "carbohydrate" recalls the deduction of early 19th-century investigators that photosynthesis made glucose directly by combining atoms of carbon with moiecules of water, as the formula for glucose suggests. In line with this idea it was thought that the oxygen transpired by green leaves came from the splitting of carbrn dioxide. The progress of chemistry, however, failed to disclose any processes that would aecomplish these results so simply. Accumulating evidance to the contrary became convincing some 30 years ago, when C.B. van Niel of Stanford University discovered that certain bacteria produce organic compounds by a process of photosynthesis similar to that in plants but with one important difference. These bacteria use hydrogen sulphide (H_2S) instead of water and liberate elemental sulphur instead of gaseous oxygen. The otherwise complete similarity of the two processes strongly suggested that the oxygen evolved by green plants must come from the splitting of water.

The Capture of Light

Photosynthesis could now be described in terms of familar chemistry. The splitting of water would be accomplished by the process of oxidatian (which means the removal of hydrogen atoms from a molecule), with oxygen gas as the product of the reaction. The free hydrogen atoms would then be available to carry through the equally familar and opposite process of reduction (which means the addition of hydrogen atoms to a molecule). By the addition of hydrogen atoms (or electrons plus hydrogen ions) the carbon dioxide would be reduced to an organic compound.

It is during the first, energy-converting stage of photosynthesis that the water molecule is split. Initially the energy of light impinging on the plant cell is transformed into the chemical potential energy of electrons excited from their normal orbits in molecules of the green pigment chlorophyll and other plant pigments. A large part of this energy eventually goes into

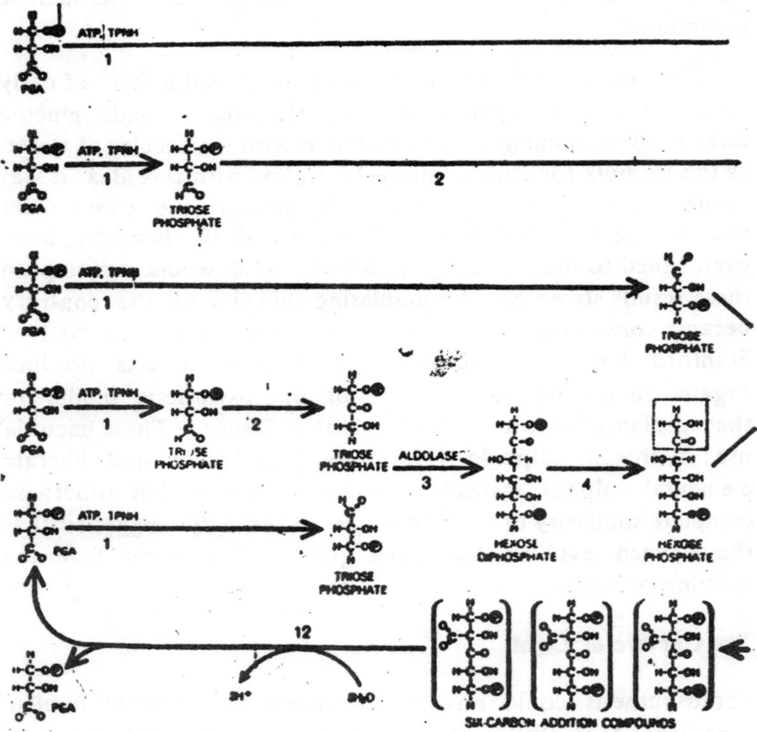

Path of Radioactive Carbon (colour) was determined from experiments described in the text. Five molecules of PGA, the first stable intermediate product to appear, are reduced (1) by cofactors ATP and TPNH to five triose phosphate molecules. A circled P represents a phosphate group ($-HPO_3^-$). Two of these are converted to a different type of triose phosphate (2); the subsequent condensation of one of each kind of triose into hexose diphosphate (3) is mediated by the enzyme aldolase. Hexose diphosphate then loses a phosphate group (4). Transketolase, another enzyme, removes two carbons from the hexose and adds them to a triose phosphate (5), making one tetrose and the splitting of water as electrons and hydrogen ions are transferred from water to the substance triphosphopyridine nucleotide (TPN^+), which is thereupon reduced to the form designated TPNH. The TPNH thus becomes not only a carrier of energy but also the bearer of electrons for the subsequent reduction of carbon dioxide. Along with the movement of

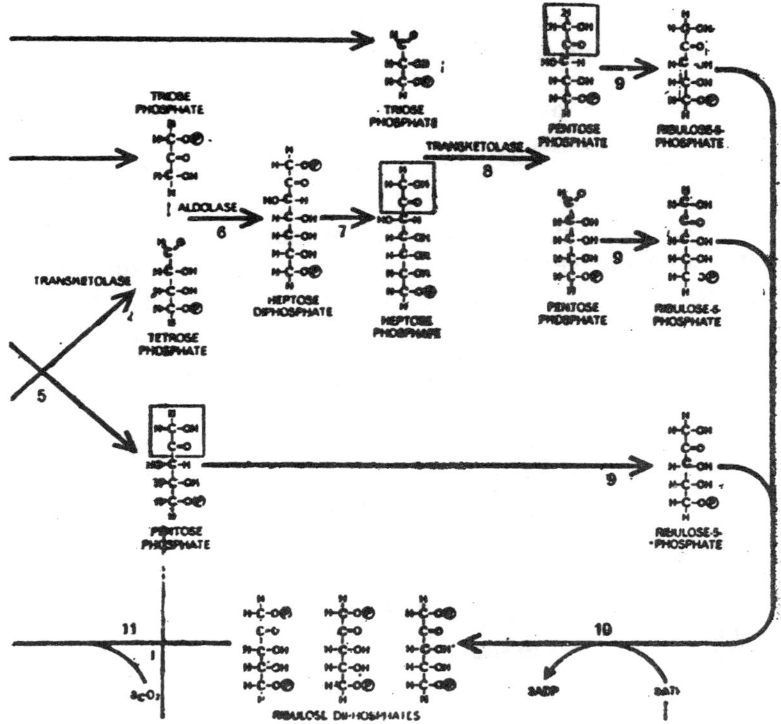

one pentose phosphate. The tetrose condenses with a triose (6) to form a heptose diphosphate, which then loses a phosphate group (7). Transketolase removes two carbons from the heptose and adds them to a triose, (8). Making two more pentose phosphates for a total of three; these are converted to ribulose 5-phosphate (9), then to ribulose diphosphate, (10). Addition of three carbon dioxide molecules, (11) produces three unstable compounds (brackets indicate unknown structure) that begin the cycle againt. Addition of three water molecules (12) results in six PGA molecules for a net gain of one in the cycle.

electrons from water to TPNH, some energy goes to charging the energy-carrying molecule adenosine triphosphate (ATP), specifically by promoting the attachment of a third phosphate group ($—OPO_3^-$) to adenosine diphosphate (ADP), the discharged form of the carrier. Both ATP and TPNH belong to the family of compounds known as cofactors or coenzymes, which work with enzymes in the catalysis of chemical reactions. ATP, the universal currency of energy transactions in the cell, plays a significant role in respiration as well as in photosynthesis.

Needless to say, the manufacture of each of these cofactors involves an intricate cycle of reactions [see "The Role of Light in Photosynthesis," by Daniel I. Arnon; Scientific American Offprint 75]. Although the cycles are not yet fully understood, it is enough for the purpose of the present discussion to know that ATP and TPNH, or closely similar compounds, furnish the energy for the second stage of photosynthesis, during which the

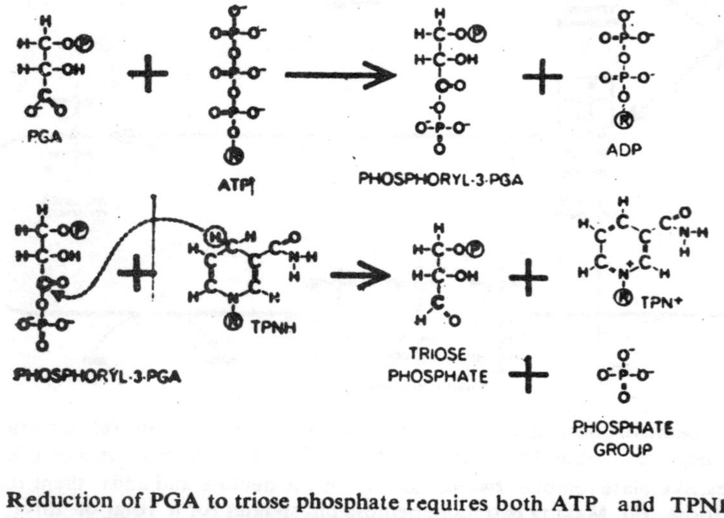

Reduction of PGA to triose phosphate requires both ATP and TPNH. At top ATP gives up its terminal phosphate group to PGA to produce phosphoryl-3-PGA. At bottom TPNH donates a hydrogen atom and an electron (broken circle and arrow), thereby displacing a phosphate group and forming triose phosphate. The second step is in reality more complex than shown here and involves other cofactors in addition to TPNH.

carbon atom of carbon dioxide is reduced and joined to a hydrogen atom and a carbon atom in place of an oxygen.

The process of reduction goes forward in small steps. Each reaction brings about some change in a carbon compound until the starting material is at last transformed to the final product. For each reaction there is therefore an intermediate compound. Since every life process involves a more or less extended series of intermediates, cells typically contain a large number of intermediates. Many of them turn up in two or more pathways leading to different end products. The tracing of the path of carbon in photosynthesis required first of all a technique for identifying the intermediates proper to it and for establishing their sequence along the path.

Samuel Ruben and Martin D. Kamen, then at the University of California, met this need some 20 years ago by their discovery of the radioactive isotope of carbon with a mass number of 14. This isotope has a conveniently long half life of more than 5,000 years; over the time period of an experiment, therefore, carbon 14 has an effectively constant radioactivity. Ruben and his colleagues recognized at once the potential usefulness of this isotope as a label for the identification of compounds in biological processes. They prepared carbon dioxide in which the carbon atoms were carbon 14. When they exposed green plants to an atmosphere containing this gas instead or normal carbon dioxide ($C^{12}O_2$), the plants took up the ($C^{14}O_2$), and made compounds from it. The presence of the carbon 14 in these compounds could be detected by various divices, such as the Geiger-Muller counter, and by radio-autography on X-ray film. Unfortunately this work was cut short by the war and by Ruben's death in a laboratory accident.

In 1946 Melvin Calvin organized a new group at the Lawrence Radiation Laboratory of the University of California with the primary objective of tracing the path of carbon in photosynthesis with $C^{14}O_2$ as one of its principal tools. Starting as a graduate student in 1947, I had the good fortune to participate in this work with Calvin, Andrew A. Benson and others

The early experiments were quite simply contrived. We used leafy plants and often just the leaves of plants. After allowing a leaf to photosynthesize for a given length of time in an atmosphere of $C^{14}O_2$ in a closed chamber, we would bring biochemical activity to a halt by immersing the leaf in alcohol. With the enzymes inactivared, the reactions converting one intermediate compound into another would story, and the pattern of labeling would be "frozen" at the point. We soon discovered, however, that photosynthesis proceeds too rapidly for completely reliable observation by such a procedure. With few seconds delay in the penetration of alcohol into the cell, for example, the labeling pattern would be disarrayed and no longer representative of the stage at which we tried to halt the photosynthesis.

Since rapid and precisely timed killing of the plant is important, we adopted single-celled algae—Chlorella pyrenoidosa and Scenedesmus obliquus—as the subject for many of our experiments. In both species the plant consists of a single cell so small that it can be seen only with a microscope. Alcohol can quickly penetrate the cell wall and deactivate the enzymes. The algae offer another advantage: they can be grown in continuous cultures, asssuring us a supply of material with highly constant properties.

An experimental sample is taken from the culture in a thin-walled, transparent closed vessel. Illuminated through the walls of the vessel and supplied wlth a stream of ordinary carbon dioxide, which is bubbled through the suspension, the algae photosynthesize at the normal rate. We shut off the supply of carbon dioxide and inject a solution of radioactive bicarbonate ion (carbon dioxide dissolved in our algae culture medium is mostly converted into bicarbonate ion). After a few seconds or minutes the cells are killed. We then extract the soluble radioactive compounds from the plant material and analyze them.

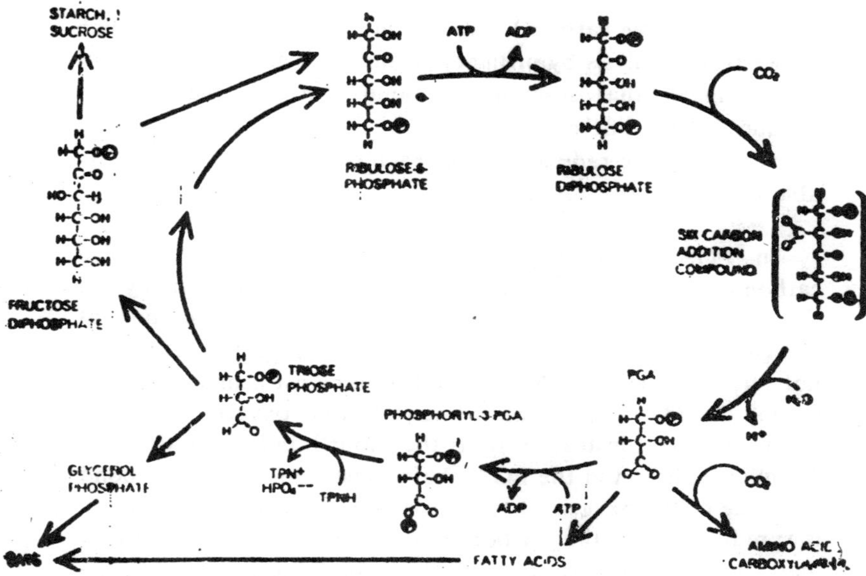

End Products of Photosynthesis are not limited to carbohydrate (*e.g.*, sucrose and starch), as first thought, but include, among other things, fatty acids, fats, carboxylic acids and amino acids. Carbon cycle shown here is highly simplified; it involves at least 12 discrete reaction. Moreover, the steps from PGA to fatty acids and to amino and carboxylic ocids have been indicated.

The Reduction of CO_2

Calvin and his colleagues soon found that the carbon 14 label was distributed among several classes of biochemical compounds, including not only sugars but also amino acids; the subunits of proteins. As the exposure time was reduced to a few seconds, the first stable intermediate product of photosynthesis was found to be the three-carbon compound 3-phosphoglyceric acid (PGA).

The next step was to determine which of the three carbon atoms in the first generation of PGA molecules synthesized in the presence of radioactive carbon dioxide bears the carbon 14 label. We first removed from PGA the phosphate group and then diluted the free glyceric acid with glyceric acid containing the stable carbon 12 isotope in order to have enough matcrial for analysis by ordinary chemical methods. Treatment with reagents that severed the bonds between the carbons produced there different products, one from each carbon atom. By measuring the radioactivity of each of the products we were able to identify the labeled carbon.

In PGA from plants that had been exposed to labeled carbon dioxide for only five seconds we found that virtually all the carbon 14 was located in the carboxyl atom. the carbon at one end of the chain that is bound to two oxygens. This was not surprising because the carboxyl group most nearly resembles carbon dioxide. The carbon is bound to the oxygens by three bonds, however, instead of four; the fourth bond now ties it to the middle carbon in the PGA chain. The transfer of this bond from one of the oxygens to a carbon constitutes the first step in the reduction of the carbon dioxide. This was evidence also that the reduction is accomplished by some sort of carboxylation reaction, a reaction in which carbon dioxide is added to some organic compound. Ultimately, of course, the two other carbons of PGA must come from carbon dioxide. But it was some time before investigation disclosed the specific compound that picks up the carbon dioxide and the cyclic pathway that makes this carbon dioxide acceptor from PGA.

The discovery of the pathway intermediates was made much easier by a then comparatively new technique; two-dimensional paper chromatography, developed by the British chemists A.J. P. Martin and R.L.M. Synge. Closely similar compounds can be distiuguished in this procedure by slight differences in their relative solubility in an organic solvent and in water. The extract from the plant is dropped on a sheet of filler paper near one-cornor. An edge of the paper adjacent to the corner is immersed in a trough containing an organic solvent; the paper is held taut by a weight and the whole assembly is placed in a water-saturated atmosphere in a vapour-tight box. The solvent travelling through the paper by capillarity dissolves the compounds and carries them along with it. As they move along in the solvent, however, the compounds tend to distribute themselves between the solvent and the water absorbed by the fibers of the paper. In general the more soluble the compared with the organic solvent, the slower it travels. If the compound is also absorbed to some extent by the cellulose fibers, its movement will be even slower. As a result the compounds are distributed in a row in one dimension. Depending on the solubility of the compounds and the nature of the solvent used, some compounds may still overlap one another. Repetition of the procedure, with a different solvent travelling at right angles to the direction of the first run, will usually separate these compounds in a second dimension.

Since most of the compounds are colourless, special techniques are needed to locate them on the paper. Those that are radioactive will locate themselves, however, if the chromatographic paper is placed in contact with a sheet of X-ray film for a few days. The resulting radio-autograph will show as many as 20 or 30 radioactive compounds in the substances extracted from algae exposed to carbon 14 for only 30 seconds [see illustration at right]. Clearly the synthetic apparatus of the plant works rapidly.

Chromatographs and Radioautographs

In order to identify these compounds we prepared a chromatographic map by running samples of many known compounds through the same chromatographic system and recording the locations at which we found them on the paper. The locations can be made visible in these cases by spraying the paper with a mist of some chemical that is known to react with the compound to produce a coloured spot. Comparison of the radioautograph of an unknwon compound with the map yields a first clue to its identity. This can be corroborated by washing the radioactive compound out of the paper with water and mixing it with a larger sample of the suspected authentic substance. The mixture is applied to a new piece of filter paper and chromatographed. With enough of the authentic material to yield a coloured spot, comparison with a radioautograph of the same paper now shows whether or not the radioactive and the authentic material really coincide. The possibility that the authentic material and the radioactive material are still not the same can be tested by using different solvent systems in the preparation of the chromatograph and by other means.

Over the years these procedures have established the identity of a great many of the intermediate and end products of photosynthesis. Some of the sugar phosphates labeled by carbon 14 proved to be well-known derivatives of triose (three-carbon) and hexose (six-carbon) sugars. Other were discovered for the first time among the intermediates produced by our algae. Benson showed that among these are a seven-carbon sugar phosphate and also five-carbon phosphates, including in particular ribulose-1, 5-diphosphate.

The rapid building of carbon 14 into the more familiar triose and hexose phosphates suggested certain biochemical pathways already established in studies of respiration. It seemed likely that PGA might be linked to these phosphates by the reverse of a sequence of respiratory reactions first mapped many years ago by the German chemists Otto Meyerhof, Gustav Embden and Jakob Parnas. In the respiratory pathway hexose phosphate is split into two molecules of triose phosphate, with the split occurring between the two carbon atoms in the middle of the chain. The triose phosphate is then oxidized to give PGA. The electrons from this energy-yielding operation are picked up by diphosphopyridine nucleotide (DPN^+), which is thereupon reduced to DPNH. The DPN^+ is a close relative of the TPN^+ that turns up in photosynthesis. In addition this oxidation yields enough energy to make a molecule of ATP from ADP and phosphate ion.

In the reverse pathway of these reactions in photosynthesis Calvin concluded, the plant uses the cofactors ATP and TPNH, made earlier by the transformation of the energy of light, to bring about the reduction of PGA to triose phosphate. In the first step the terminal phosphate group of ATP is transferred to the carboxyl group of PGA to form a "carboxyl phosphate" (really an acyl phosphate). Some of the chemical potential energy that was stored in ATP is now stored in the acyl phosphate, making the new intermediate compound highly reactive. It is now ready for reduction by TPNH. This reducing agent donates two electrons to the reactive intermediate. One carbon-oxygen bond is thereby severed and the oxygen atom, carried off with the phosphate group, is replaced by a hydrogen atom. In this way the carboxyl carbon atom is reduced to an aldehyde carbon atom; that is, it now has two bonds to oxygen instead of three and one bond each to carbon and hydrogen. (see illustration on page 104]. This is the point in the cycle at which most of the solar energy captured in the first stage of photosynthesis is applied to the reduction of carbon.

The Unstable Intermediate

The next development in the plotting of the carbon pathway came from a series of experiments first performed in our laboratory by Peter Massini. He hoped to see which intermediates would be most trongly increased or decreased in concentration by turning off the light and allowing the synthetic process to go on for a while in the dark. In order to establish the concentration of the various intermediates when the reaction proceeds in the light, he bubbled radioactive carbon dioxide through the culture for more than half an hour. At the end of this period every intermediate was as highly radioactive as the incoming carbon dioxide. The radioactivity from each compound therefore gave a measure of the concentration of the compound. He then turned off the light and after a few seconds took another sample of algae in which he measured the relative concentration of compounds by the same technique. Comparison with the compounds sampled in the light showed that the concentration of PGA was greatly increased. This finding could be readily explained: turning off the light stopped the production of the ATP and TPNH required to reduce PGA to triose phosphate.

Of the sugar phosphates present, only one, the five-carbon

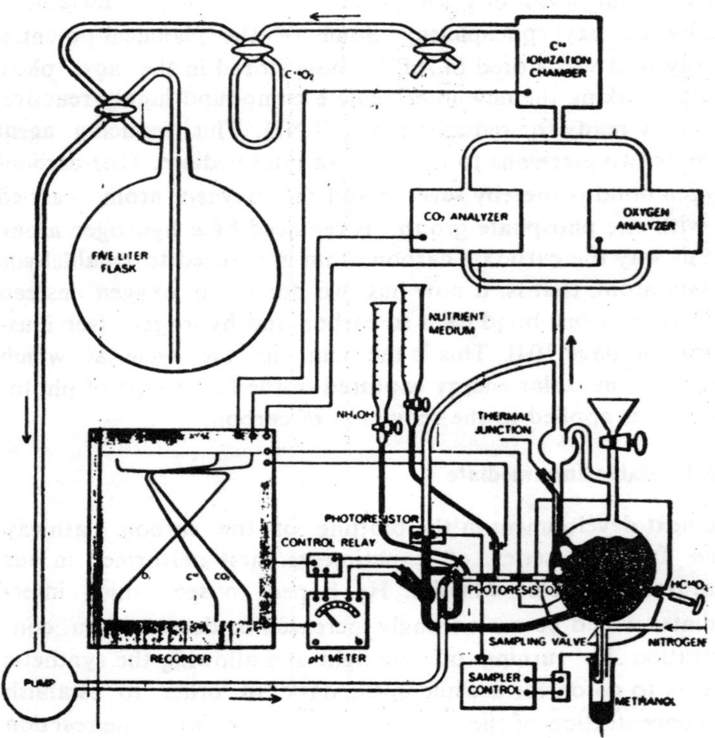

Fig. 9.1: Steady-State Apparatus permits experimental control and study of photosynthesis. The algae are suspended in nutrient in a transparent vessel (*lower right*). A gas pump circulates a mixture of air, ordinary carbon dioxide and labeled carbon dioxide (when needed) to the vessel, where it bubbles through the suspensions. Labeled carbon can also be added in the form of bicarbonate ($HC^{14}O_3$). Measurements of the oxygen, carbon dioxide and labeled carbon levels in the gas are recorded continuously. The pH is maintained at a constant value by means of the pH meter. The sampler control allows removal of samples into the test tube.

ribulose diphosphate, was found to have changed significantly; its concentration dropped to zero. Because the PGA had simultaneously increased in concentration, it was apparent that ribulose diphosphate was consumed in the production of PGA.

This finding was of great significance because it indicated for the first time that ribulose diphosphate is the intermediate to which carbon dioxide is attached by the carboxylation reaction.

For this reaction ribulose diphosphate is prepared by an earlier reaction that goes on in the light and in which ATP donates its terminal phosphate group to ribulose monophosphate. The more reactive diphosphate molecule now adds one molecule of carbon dioxide by carboxylation. The details of this reaction remain obscure because the resulting six-carbon intermediate is so unstable that we have not been able to detect it by our methods of analysis. As its first stable product this sequence of events yields two three-carbon PGA molecules.

Massini's experimental results were confirmed by a parallel experiment devised by Alex Wilson, then a graduate student in our laboratory. Instead of turning out the light Wilson shut off the supply of carbon dioxide. In this situation one might expect to find an increase in the concentration of the compound that is consumed in the carboxylation reaction; ribulose diphosphate showed such an increase. Corrrespondingly, one would look for a decrease in the concentration of the product of this reaction; PGA did in fact decrease in concentration.

The first steps along the path were thus established. The photosynthesizing plant starts with ribulsose monophosphate and converts it to ribulose diphosphate, using chemical potential energy trapped from the light in the terminal phosphosphate bond of ATP. Carbon dioxide is joined to this compound, and the resulting six-carbon intermediate splits to two molecules of PGA. With energy and electrons supplied by ATP and TPNH, PGA is reduced to triose phosphate. In the next step, it was apparent, two triose phosphates must be joined end to end in the reverse of a familiar respiratory pathway to form a hexose phosphate. The pathway from hexose to pentose phosphate remained to be uncovered.

We continued the carbon-by-corbon dissection and analysis

of these chains by the methods that had earlier shown the carbon 14 in PGA to be located first in the carboxyl carbon. In the hexose molecules we had found the labeled carbon concentrated in the two middle carbons, just where it should be if two triose molecules mode from PGA were linked together by their labeled ends. We also took apart the seven-carbon and five-carbon sugar phosphates to establish the position of the carbon 14 atoms in their chains. As the result of these degradations we were able to show that the over-all economy of the photosynthetic process starts with five three-carbon PGA's, variously transforms them through three-, six-four-and seven-carbon phosphates intermediates and returns three five-carbon ribulose diphosphates to the starting point [see illustrations on pages 192 and 193]. From carboxylation of these three chains and their immediate bisection, the cycle at last yields six PGA molecules. The net result, therefore, is the conversion of three carbon dioxide molecules to one PGA molecule.

The Calvin Cycle

With these steps filled in, the carbon reduction cycle in photosynthesis, called the Calvin cycle, was complete. The intermediates formed in the cycle depart from it on various pathways to be converted to the end products of photosynthesis. From triose phosphate, for example, one sequence of reactions leads to the six-caabon sugar glucose and the large family of carbohydrates.

Because the cycle had been established primarily by experiments with algae and the leaves of a few higher plants, it was important to see whether or not the cyele prevailed throughout the plant kingdom. Calvin and Louisa and Richard Norris carried out experiments with a wide variety of photosynthetic organisms. In every case, although they found variation in the amounts of particular intermediates formed, the pattern was qualitatively the same.

It also had to be shown that the path-way we had traced

1. Hexose and Heptose Monophosphates
2. Pga
3. Sucrose
4. Hexose and Heptose Diphosphates

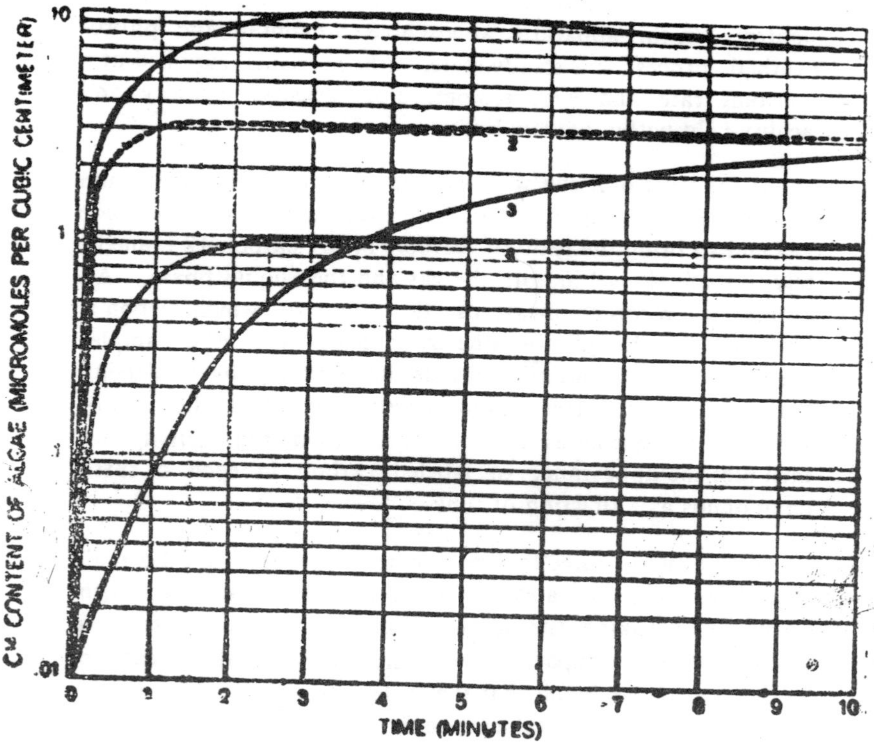

Fig. 9.2: Calvin Cycle (see illustration on page 192 and 193) was proven the most important route of carbon reduction in photosynthesis in studies by the author and Martha Kirk. As seen here, all stable intermediates of the cycle become saturated with labeled carbon within three to five minutes. The rate of carbon uptake in the algae culture showed the cyele accunts for more than 70 per cent of the carbon fixed in compounds.

out is quantitatively the most important route of carbon reduction in photosynthesis. To this end Martha Kirk and I undertook an intensive study of the kinetics of the flow of carbon in photosynthesis. Our study has helped to solve other general problems, particularly the question of how carbon enters into the pathways leading to the synthesis of proteins and fats. The biological materials for this work are supplied by an algae culture system under automatic feedback control. In this apparatus we are able to maintain the photosynthetic process in a steady state, with nutrients supplied at a constant rate and with temperature, density, salinity and acidity held within narrow limits.

At the start of a run we inject radio-active bicarbonate ion into the culture medium along with radioactive carbon dioxide gas and so bring the ratio of carbon 14 to carbon 12 immeditely to its final level in both the gas and the liquid phase. An automatic recorder measures the rate at which carbon is absorbed by the photosynthesizing cells. We take samples every few seconds and kill the cells immediately by immersing them in alcohol. After we have chromatographed the photosynthetic intermediates and measured their radioactivity we then plot the appearance of labeled carbon in each of these compounds as a fu iction of time.

By the end of three to five minutes, our records show, all the stable inter-mediates of the cycles are saturated with carbon 14. Taking the total amount of carbon thus fixed in compounds and comparing it with the rate of uptake of carbon in the culture, we found that the cycle accounts for more than 70 per cent of the total carbon fixed by the algae. A small but significant amount is also taken up by the addition of carbon dioxide to a three-carbon compound, phosphoenolpyruvic acid, to give four-carbon compounds

From the earliest work with carbon 14 in our laboratory, it had been apparent that carbon dioxide finds its way rather quickly into products other than carbohydrates in the photosynthesizing plant. This was at variance with traditional ideas about photosoynthesis that regarded carbohydrates as the sole organic products of the process. It was important to ask, therefore, whether fats and amino acids could be formed directly from the cycle as products of its intermediates or whether these noncarbohydrates were synthesized only from the carbohydrate end products of photosyntesis. Our kinetic studies show that certain amino acids must indeed be formed from the intermediates and must therefore be regarded as true products of photosynthesis. The amino acid alanine, for example, shows up labeled by carbon 14 at least as rapidly as any carbohydrate; it would be labeled with carbon 14 much more slowly if it were made from carbohydrate, since the carbohydrate would have to be labeled first. We have been able to show that more than 30 per cent of the carbon taken up by the algae in our steady-state system is incorporated directly into amino acids. There is some evidence that fats may also be formed as products of the cycle.

The discovery that plants make these other compounds as direct products of photosynthesis lends new interest and importance to the chloroplast, the subcellular compartment of green cell that contains pigments and the rest of the photosynthetic apparatus. It has been known for some time that this highly structured organelle is responsible for the absorption of light, the splitting of water and the formation of the cofactors for carbon reduction. More recent studies have shown that it is the site of the entire carbon-reduction cycle. Now the chloroplast emerges as a complete photosynthetic factory for the production of just about everything necessary to plant's growth and function.

10

The Ecosphere

The great 19th-century French naturalist Jean Lamarck first conceived the idea of the biosphere as the collective totality of living creatures on the earth, and the concept has been taken up and developed in recent years by the Russian geochemist V.I. Vernadsky. The word "ecosystem" means a self-sustaining community of organisms—plants as well as animals—taken together with its inorganic environment.

Now all these are interdependent. Animal life could not exist without plants nor plants without animals, which supply them with carbon dioxide. Even the composition of the inorganic environment depends upon the cyclic activity of life. Photo synthesis by the earth's plants would remove all of the carbon dioxide from the atmosphere within a year or so if it were not returned by fires and by the respiration of animals and other consumers of plants. Similarly nitrogen-fixing organisms would exhaust all of the nitrogen in the air in less than a million years. And so on. The conclusion is that a self-sustaining community must contain not just plants, animals and nitrogen-fixers but also decomposers which can free the chemicals bound in protoplasm. It is very fortunate from our standpoint that some microorganisms have solved the biochemical trick of decomposing chitin, lignins and other inert organic compounds that tie up carbon.

A community must consist of producers or accumulators of energy (green plants), primary consumers (fungi, microorganisms and herbivores), higher-order consumers (carnivorous predators, parasites and scavengers), and decomposers that regenerate the raw materials.

Organisms living on the face of the earth as it floats around in space can receive energy from several sources. Energy from outside comes to us as sunlight and starlight, is reflected to us as moonlight, and is brought to earth by cosmic radiation and meteors. Internally the earth is heated by radioactivity and it is also gaining heat energy from the tidal friction that is gradually slowing our rotation. On top of this man is tapping enormous amounts of stored energy by burning fossil fuels. But all these secondary sources of energy are infinitesimal compared to our daily sunshine, which accounts for 99.9998 per cent of our total energy income.

This supply of solar energy amounts to 13×10^{28} gram-calories per year, or, if you prefer, it represents a continuous power supply at the rate of 2.5 billion billiom horsepower. About onethird of the incoming energy is lost at once by being reflected back to space, chiefly by clouds. The rest is absorbed by the atmosphere and the earth itself, to remain here temporarily until it is re-radiated to space as heat. During its residence on earth this energy serves to melt lice, to warm the land and oceans, to evaporate water, to generate winds and waves and currents. In addition to these activities, a ridiculously small proportion—about four hundredths of 1 per cent—of the solar energy goes to feed the metabolism of the biosphere.

Practically all of this energy enters the biosphere by means of photosynthesis. The plants use one-sixth of the energy they take up from sunlight for their own metabolism, making the other five-sixths available for animals and other consumers.

About 5 per cent of this net energy is dissipated by forest and grass fires and by man's burning of plant products as fuel.

When an animal or other consumer eats plant protoplasm, it uses some of the substance for energy to fuel its metabolism and some as raw materials for growth. Some it discharges in brokendown form as metabolic waste products: for example, animals excrete urea, and yeast releases ethyl alcohol. And a large part of the plant material it ingests is simply indigestible and passes through the body unused. Herbivores, whether they are insects, rabbits, geese or cattle, succeed in extracting only about 50 per cent of the calories stored in the plant protoplasm. (The lost calories are, however, extractable by other consumers: flies may feed on the excretions or man himself may burn cattle dung for fuel.)

Of the plant calories consumed by an animal that eats the plant, only 20 to 30 per cent is actually built into protoplasm. Thus, since half of its consumption is lost as waste, the net efficiency of a herbivore in converting plant protoplasm into meat is about 10 to 15 per cent. The secondary consumers—*i.e.*, meat-eaters feeding on the herbivores—do a little better. Because animal protoplasm has a smaller proportion of indigestible matter than plants have, a carnivore can use 70 per cent of the meat for its internal chemistry. But again only 30 per cent at most goes into building tissue. So the maximum efficiency of carnivores in converting one kind of meat into another is 20 per cent.

Some of the consequences of these relationships are of general interest and are fairly well known. For example, 1,000 calories stored up by the algae in Cayuga Lake can be converted into protoplasm amounting to 150 calories by small aquatic animals. In turn, smelt eating these animals produce 30 calories of protoplasm from the 150. If a man then eats the smelt he can synthesize six calories worth of fat or muscle from the 30; if he waits for the smelt to be eaten by a trout and then eats the trout, the yield shrinks to 1.2 calories. If we were really dependent on the lake for food, we would do well to exterminate the trout and eat the smelt ourselves, or, better yet, to exterminate the smelt and live on planktonburgers. The same principles, of course, apply on land. If man is really determined to support the largest possible populations of his kind, he will have to shorten the food chains leading to himself and, so far as practicable, turn to a vegetarian diet.

The rapid shrinkage of stored energy as it passes from one organism to another serves to make the study of natural communities a trifle more simple for the ecologist than it would otherwise be. It explains why food chains in nature rarely contain more than four or five links. Thus in our Cayuga Lake chain the trout was the third animal link and man the fourth. Chains of the same sort occur in the ocean, with, for example, a tuna or cod as the third link and perhaps a shark or a seal replacing man as the fourth link. Now if we look for the fifth link in the chain we find that it takes something like a killer whale or a polar bear to be able to subsist on seals. As to a sixth link—it would take quite a predator to make its living by devouring killer whales or polar bears.

We could, of course, trace food chains in other directions. Each species has its parasites that extort their cut of the stored energy, and these in turn support other parasites down to the point where there is not enough energy available to support another organism. Also, we should not forget the unused energy contained in the feces and urine of each animal. The organic matter in feces is often the basic resource of a food chain in which the next link may be a dung beetle or the larva of a fly.

It is estimated that the maximum amount of protoplasm of all types that can be produced on earth each year amounts to 410 billion tons, of which 290 billion represent plant growth and the other 120 billion all of the consumer organisms. We see, then, that the availability of energy sets a limit to the amount of life on earth—that is, to the size of the biosphere. This energy also keeps the nonliving part of the ecosphere animated, largely through the agency of moving water, which is the single most important chemical substance in the physiology of the ecosphere.

Each year the oceans evaporate a quantity of water equivalent to an average depth of one meter. The total evaporatian from land and bodies of fresh water is one sixth of the evaporation from the sea, and at least one fifth of this evaporation is from the transpiration of plants growing on land. The grand total of water evaporated annually is roughly 100,000 cubic miles, and this must be roughly the annual precipitation. The precipitation on land exceeds the evaporation by slightly over 9,000 cubic rules, which therefore represents the annual runoff of water from land to sea. It is astonishing to me to note that more than one tenth of this total runoff is carried to the sea by just two rivers-the Amazon and the Congo.

Precipitation supplies nonmarine organisms with the water which they require in large quantities. Protoplasm averages at least 75 per cent water, and plants require something like 450 grams of water to produce one gram of dry organic matter. The water moving from land to sea also erodes the land surface and dissolves soluble mineral matter. It brings to the plants the chemical nutrients that they require and it tends to level the land surface and deposit the minerals in the sea. At present the continents are being worm down at an average world-wide rate of one centimeter per century. The levelling process, however, apparently has never gone on to completion on the earth. Geological uplift of the land always intervenes and brings marine sediments above sea level, where the cycle can begin again.

The rivers of the world are now washing into the seas some four billion tons of dissolved inorganic matter a year, about 400 million tons of dissolved organic matter and about five times as much undissolved matter. The undissolved matter represents destruction of the land where organisms live, but the dissolved materials is of greater interest, because it includes such important chemicals as 3.5 million tons of phosphorus, 100 million tons of potassium and 10 million tons of fixed nitrogen. In order to say what these losses may mean to the biosphere we must review a few facts about the chemical composition of the earth and of organisms.

Every organism seems to require at least 20 chemical elements and probably several others in trace amounts. Some of the organisms' requirements are rather surprising. Penicillium is said to need traces of tungsten, and the common duckweed demands manganese and the rare earth gallium. There is a European pansy which needs high concentrations of zinc in the soil, and several plants in different parts of the world are so hungry for copper that they help prospectors to find the mineral. Many organisms have fantastic abilities to concentrate the necessary elements from dilute media. The sea-squirts have vanadium in their blood, and the liver of the edible scallop contains on a dry-weight basis one tenth of 1 per cent of cadmium, although the amount of this element in sea water is so small that it cannot be detected by chemical tests.

But the exotic chemical tastes of organisms are compratively unimportant. Their main needs can be summed up in just five words—oxygen, carbon, hydrogen, nitrogen and phosphorus, which account for more than 95 per cent of the mass of all protoplasm. Oxygen is the most abundant chemical element on earth, so we probably do not need to be concerned about any absolute deficiency of oxygen. But nitrogen is a different matter. Whereas protein, the main stuff of life, is 18 per cent nitrogen, the relative abundance of this element on the earth is only one 10,000th of the earth's mass. It is apparent that our land forms of life could not long tolerate a net annual loss of 10 million tons of fixed nitrogen to the sea. Fortunately this nitrogen loss from land is reversible, so that we can speak of a "nitrogen cycle." Organisms in the sea convert the fixed nitrogen into ammonia, a gas which can return to land via the atmosphere.

Carbon also is not in too abundant supply, for it amounts to less than three parts in 10,000 of the total mass of the earth's matter. But once again the biosphere profits from the fact that carbon can escape from the oceans as a carbon dioxide. This gas goes through a complex circulation in the atmosphere,

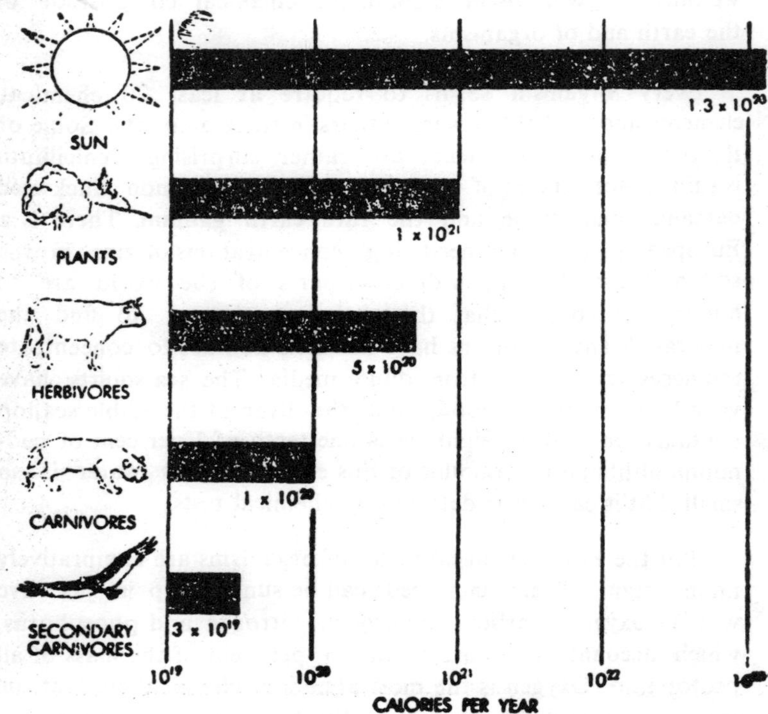

Fig. 10.1: Utilization of Solar Energy decreases with each step along the food chain. These bars (on a logarithmic scale) show that plants use only .08 per cent of energy reaching the atmosphere; plant-eaters use only part of this fraction and flesh-eaters even less.

being released from the oceans in tropical regions and absorbed by the ocean waters in polar regions. Because some carbon is deposited in ocean sediments as carbonates, there is a net loss of carbon from the ecosphere. But there seems to be no danger that a shortage of this element will restrict life. The atmosphere contains 2,400 billion tons of carbon dioxide, and at least 30 times that much is dissolved in the oceans, waiting to be released if the atmosphere should become depleted. Volcanoes discharge carbon dioxide, and man is burning fossil fuels at such a rate that he has been accused of increasing the average carbon dioxide content of the atmosphere by some 10 per cent in the last 50 years. In addition, lots of limestone, which is more than 4 per cent carbon dioxide, has been pushed up from ancient seas by uplifts of the earth.

The story of phosphorus appears somewhat more alarming. This element accounts for a bit more than one tenth of 1 per cent of the mass of terrestrial matter, is enriched to about twice this level in plant protoplasm and is greatly enriched in animals, accounting for more than 1 per cent of the weight of the human body. As a constituent of nucleic acids it is indispensable for

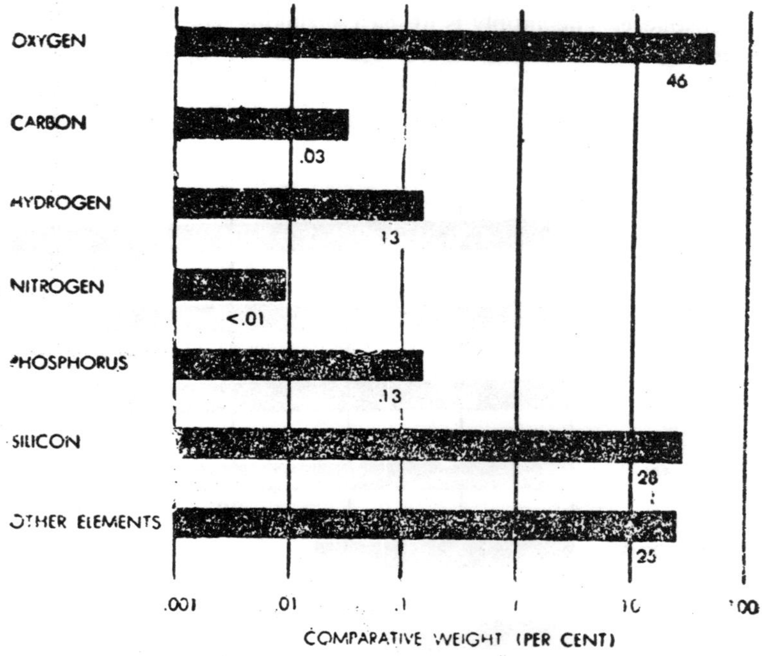

Fig. 10.2: Estimated Relative Abundance of elements in the earth and its atmosphere (*above*) and in living matter (*below*) is compared in these charts; the scale is logarithmic. Silicon, with many stable compounds, is abundant on earth but rare in living organisms. Nitrogen, rare, on earth is important to life, making up as much as 18 per cent of proteins.

all types of life known to us. But many agricultural lands already suffer a deficiency of phosphorus, and a corn crop of 60 bushels per acre removes 10 per cent of the phosphorus in the upper six inches of fertile soil. Each year 3.5 million tons of phosphorus are washed from the land and precipitated in the seas. And unfortunately phosphorus does not escape from the sea as a gas. Its only important recovery from the sea is in the guano produced by sea birds, but less than 3 per cent of the phosphorus annually lost from the land is returned in this way.

It must agree with agriculturalists who say that phosphorus is the critical limiting resource for the functioning of the ecosphere. The supply is at least shrinking (if dwindling is too strong a word) end there seems to be no practical way of improving the situation short of waiting for the next geological cycie of uplift to bring phosphate rock above sea

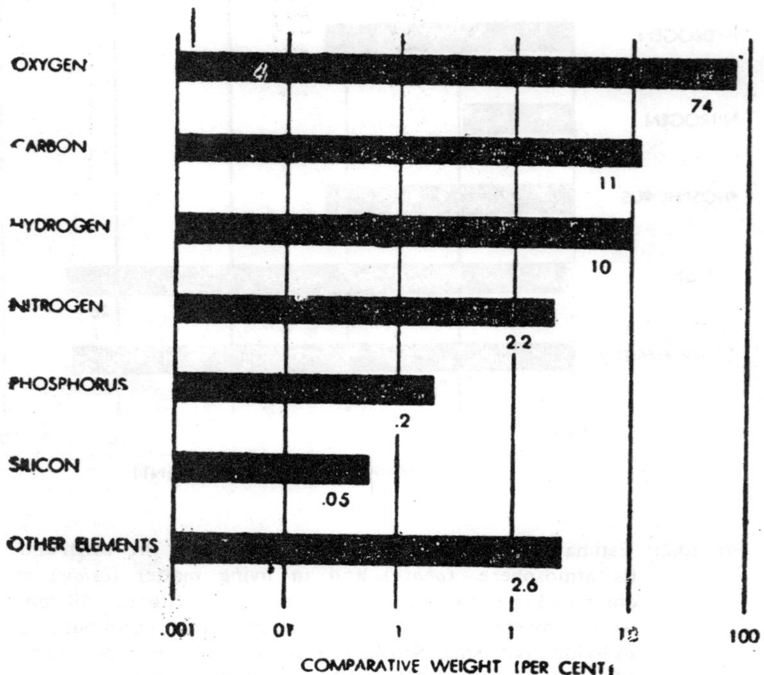

level. Perhaps we should also worry about other essential elements, such as calcium, potassium, magnesium and iron, which behave much like phosphorus in the metabolism of the ecosphere, but the evidence clearly indicates that if present trends continue phosphorus will be the first to run out.

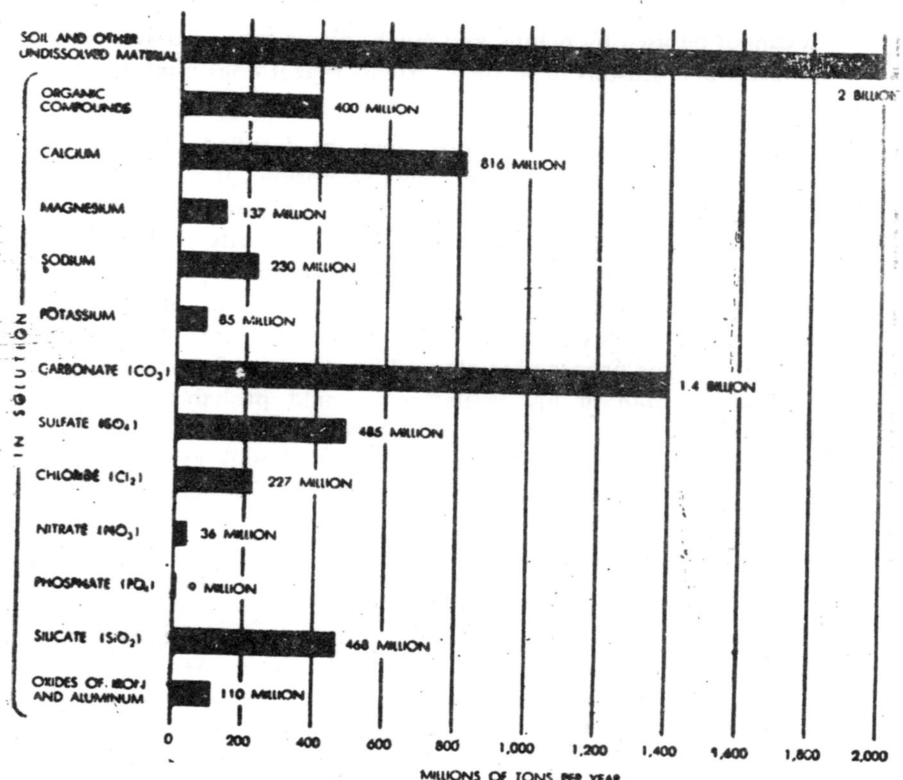

Fig. 10.3. Annual Loss of minerals and organic matter washed into the sea amounts to billions of ions. Much nitrogen and carbon eventually return to the land via the atmosphere; the loss of phosphate is more serious since almost all of it remains in the oceans.

This brings me to the close of a very superficial summary of dome of the physiological processes of the ecosphere. There are srastic oversimplifications in this treatment; the importance of some processes may be overestimated and others (*e.g.*, dumping sewage in rivers and oceans) may not have received enough attention. The figures for the total quantity of energy received by the earth, for total annual precipitation and for the total supply of some chemical elements may overlook the very irregular distribution of these resources in time and space. Much solar energy falls on deserts and fields of snow and ice where it cannot be used by plants, and much precipitation arrives at unfavourable seasons or in such torrents that it does more harm than good to organisms.

Our survey suggests that man may be justified in feeling some real concern about the problem of erosion. It should also make us aware of the important role played by organisms that we might otherwise ignore or even regard as pests. The dung beetles, the various scavengers and the termites and other decomposers all play important bit parts in this great production. At least six diverse groups of bacteria are absolutely essential for the proper physiological functioning of the nitrogen cycle alone. Man in his carelessness would probably neither notice nor care if by some unlikely chance his radioactive fall-out or one of his chemical sprays or fumes should exterminate all of the microorganisms that are capable of decomposing chitin. Yet, as we have seen, such a tragedy would eventually mean an end to life on earth.

Finally, it is interesting to ask how large a role man plays in the physiology of the ecosphere. The Statistical Office of the United Nations estimates the present human population of the earth at 2.7 billion persons. Each of these is supposed to consume at least 2,200 metabolizable kilocalories per day. This makes a total food requirement of 22×10^{14} kilocalories per year. I have estimated that all of the plant growth in the world amounts to an annual net of all 5×10^{17} kilocalories, of which not more than 50 per cent is metabolizable by any primary consumer. Thus if man were to feed exclusively on plants he would require almost exactly 1 per cent of the total productivity of the earth.

To me [illegible] is a very impressive figure. There are more than one million species of animals, and when just one of these million species can corner 1 per cent of the total food resources, this form is truly in a position of overwhelming dominance. The figure becomes even more impressive when we reflect that 70 per cent of the total plant production takes place in the oceans, and that our figure for productivity includes inedible materials such as straw and lumber.

11

Ecological Niches

In this chapter a close look at species packing in parasite communities will be undertaken in an attempt to resolve the divergence of opinion. Attention is limited to the literature on helminth parasites of vertebrates because it illustrates the full range of types of coexistence seen in any taxon and it contains some of the most celebrated examples of coexistence of parasites, some incredibly rich communities within one organ system, excellent experimental work revaling the fundamental and realized niches (Hutchinson, 1957) of parasities and is deserving of much more attention by ecologists. In the review by Schoener (1974) on resource partitioning in ecological communities covering 80 studies, none on helminths is included.

Some studies are summarized in Table 11.1 with a statement on the predominant type of coexistence in the community. The kind of evidence provided in the studies is also listed with a check indicating quantitiative data or experimental manipulation.

The general pattern of coexistence is clear (Table 11.1). The vast majority of cases indicates a predominantly non-interactive niche occupation (no substantial shift from fundamental to realized niche in presence of another species), either in nonoverlapping niches or in strongly overlapping niches. In the latter case all species occur in the alimentary canal and are usually members of very rich helminth faunas. Although studies have not gone far in determining how these species coexist while overlapping so broadly along the length of the alimentary canal, it is clear that other niche dimensions must be considered. Schad (1963a, b) found in the genus Tachygonetria in the tortoise gut that some species occur near the gut mucosa while other can be found throughout the gut lumen. Some species feed on large particle sizes, others specialize on bacteria, and others imbibe purely liquid food. Given the great physico-chemical diversity of the gut, many other parameters may be involved in the niche diversification in these gut parasites (See MacKenzie and Gibson, 1970; Williams, McVicar, and Ralph, 1970). Thus the group of studies showing predominantly overlapping but noninteractive niche occupation illustrate overlap only on the most obvious dimension of gut length, which has been the most commonly studied. Since niche divergence on other gradients is evident, this group of studies should be classed with the other non-interactive examples making 18 studies out of the tot al of this type. In the nonoverlapping class some studies have been placed that show seemingly insignificant amounts of overlap (30 per cent or less), again making the distinction an arbitrary one.

Holmes (1973) also found many examples of what he called selective site segregation of niches where niche occupation was uninfluenced by the presence of other parasite species. As in Table 11.1 he supplied many examples of niche segregation of this type and only few where coexistence involved reduced realized niches in the presence of competitors, *i.e.*, interactive site segregation. He concluded that the majority of parasite communities are mature because the residents have evolved discrete niches without competition. That is, communities have reached the evolutionary phase in Wilson's (1969) concept of community development and have presumably passed the earlier noninteractive, interactive, and assortative phases.

TABLE 11.1

Some Studies Illustrating Coexistence of Helminth Parasites in Vertebrates

Genus of Parasite	Number of Species	Host and Organs Infeted	Reference	Quant. data	Expermtl. manip.	Type of coexistence
1	*2*	*3*	*4*	*5*	*6*	*7*
1. Strongyloids	2	Rat, alimentar canal	Wertheim (1970)	√	√	Non overlapping, Non interactive
2. Calycotyle	2	Weasel, proctodeal region	Euzet & Williams (1960)	—	—	,, ,,
3. Aporocotyle & Psettarium	2	Rockfish spp., heart and bronchial system	Holmes (1971)	√	—	,, ,,
4. Kalicephalus	3	Racer (snake) alimentary canal	Schad (1956, 1926b)	—	—	,, ,,
5. Filaroides & others	3	Mink, lung	Stockdale (1970)	—	—	,, ,,
6. Apocreadium	3	Ocean tally (fish) alimentary canal	Sogandares-Bernal (1959)	—	—	,, ,,
7. Castroia	2	Bat, alimentary canal	Martin (1969)	√	—	,, ,,

8.	Proteocephalus & Neoechinorhynchus*	2	Cisco (fish) alimentary canal	Cross (1934)	√	—	Non overlapping,	Interactive
9.	Tachygonetria	8	Tortoisc, alimentary canal	Schad (1962a, 1963a, b)	√	—	Overlapping,	Non interactiye
10.	Tachygonetria & others	14	Tortoisc, alimentary canal	Petter (1966)	√	—	,,	,,
11.	Cylicocyclus & other strongylids	32	Horse, alimentary canal	Foster (1936)	√	—	,,	,,
12.	Hymenolepis & others*	43	Scaup, alimentary canal	Hair & Holmes (1975)	√	—	,,	,,
13.	Crepidostomum & others*	6	Trout, most in alimentary canal	Thomas (1964)	√	—	,,	,,
14.	Cucullanus & others*	11	Flounder, alimentary canal	MacKenzic & Gibson (1970)	√	—	,,	,,
15.	Capillaria & others	9	Ray, alimentary canal	Williams, McVicar, & Ralph (1970)	—	—	,,	,,
16.	Lepidapedon others	15	Cod, alimentary canal	Williams, McVicar, & Ralph (1970)	—	—	,,	,,

(Contd.)

TABLE 11.1 (*Contd.*)

1	2	*3*	*4*	*5*	*6*	7	
17. Hymenolepis & Moniliformes*	2	Hamster, alimentary canal	Holmes (1962b)	√	√	,,	,,
18. Echinor-hynchus*	>1	Fish spp., alimentary canal	Chubb (1964)	—	—	,,	,,
19. Itygonimus & Omphalometra	3+	Mole, alimentary canal	Frankland (1959)	√	—	,,	,,
20. Hymenoplepis & Moniliformes*	2	Rat, alimentary canal	Holmes (1961, 1962a)	√	√	Overlapping, niche shifts	Interactive,
21. Proteocephalus & Neoechi-norhynchus*	2	Stickleback, alimentary canal	Chappell (1969)	√	—	,,	,,
22. Dactylogyrus	4	Carp, gills	Paperna (1964)	√	√	,,	,, exclusion

*Indicates presence of an acanthocephalan in the community.

The fact that there are two phases in community development with predominantly noninteractive niche occupation, and these at the two extremes of community age, should make us pause before reaching conclusions from the observation that many parasite communities are predominantly noninteractive. The concepts stressed in this monograph favour the conclusion that parasite communities are in an early stage of development. Noninteractive niche exploitation is common because resources remain unutilized since species have not evolved to use them. Communities are largely in the noninteractive and interactive phases of development and seldom have the assortative and evolutionary phases been reached. Each species is specialized to a narrow range of resources because of coevolutionary demands other than competition. A closer look at the helminth community literature, which follows in the next two sections of this chapter, justifies the latter conclusion in my opioion.

Of course, competition can be observed among helminth gut parasites (e.g., see studies 20 and 21 in Table 11.1), but I contend that it affects a minority of species. The influence of immune responses may also prove to be important, but, again, evidence is wanting that large numbers of species interact via the host's immune system. Crompton (1973) reviewed the literature on 252 species of helminth in the alimentary canal of vertebrates, but he cited no clear examples of interspecific interaction through immunological responses and only four examples of niche shifts through interspecific competition (including Holmes, 1961; and Chappell, 1969 in Table 11.1). But eight examples were provided for the extension of niche breadth in response to intraspecific competition as parasite populations increased in a host, suggesting that interspecific constraints on niche expansion were absent. Particularly in the gut, but perhaps elsewhere also, resources may be replenished by a host so rapidly that they are not limiting except at the most extreme parasite densities (when the host is likely to die with the consequent death of the parasites).

NONINTERACTIVE COEXISTENCE

There is little evidence to suggest that competition has been an organizing force in the studies indicating nonoverlapping, non-interactive coexistence of parasites (studies 1 to 7 in Table 11.1). In mink pulmonary tissues, Stockdale (1970) reports three species of metastrongyle nematode each with a very distinct niche Perostrongylus pridhami in alveolar ducts and terminal and respiratory bronchioles; Filaroides marits in peribronchial connective tissue; Crenosoma hermani in bronchi. Stockdale notes that pulmonary tissues provide a diverse array of resources available to nematodes: alveoli and alveolar ducts, terminal and respiratory bronchioles, bronchi, lamina propria of the bronchial bifurcation, peribronchial connective tissue and pulmonary arteries. In a range of animals all these sites have been colonized but in the mink little more than three of the six sites are utilized. The nematodes are loosely packed, almost twice as many species could coexist in the pulmonary community, and it is most unlikely that competition could cause such complete divergence of ecological niches.

Two species of blood flukes in rockfishes had fundamental niches that showed a similarity of 7 per cent and when together this similarity was reduced by only 3 per eent (see data in Holmes, 1971). Indeed, the species could not overlap significantly because they are adapted in fundamentally different ways to lodge in different parts of the vascular system. Aporocotyle macfarlani is a squat fluke that wedges into the walls of blood vessels, largely in the gill arches. Psettarium sebastodorum is a longer, narrower fluke that loops its body and is so held largely in the chambers of the heart. These species have a grossly different body form that preadapted them for coexistence. Divergence of niches once both species had colonized that host was probably miniuscule. Similar examples are described by Llewellyn (1956), Williams (1960), and Uglem and Beck (1972).

In the snake Coluber constrictor three species of Kalicephalus can be found: K. inermis in the anterior part of the esophagus, K. costatus in the duodenum, and K. rectiphilus in the rectum (Schad, 1956, 1962b). Species packing is so loose it could not be created by competitive interaction but rather by the demands for specialization, leaving resources available for other species to colonize.

Three species of Apocreadium occupy largely nonoverlapping niches: A. balistis the first third of the intestine, A. uroproctoferum the second third, and A. coili the rectum. Had competition been a force in organization we should expect more overlap even to the extent that some species have Hutchinsonian niche differences of only 20 to 30 per cent (see Hutchinson, 1959).

In the literature describing species with largely overlapping niches the evidence for competition commonly acting as an organizing force is also very unimpressive. Schad (1963b) claimed that the pinworm communities in tortoises showed non-overlapping niche occupation with the inference of tight species packing. But this claim was based on correspondence of relative abundances to Mac-Arthur's (1957) broken stick model that can be derived from very different assumptions and is thus discredited as a means of distinguishing organizational influences in a community (Pielou, 1969; Poole, 1974). Based on Schad's (1962a, 1963a, b) studies, we could reasonably expect in the tortoise gut a group of species specialized to each of at least four sections of the gut, each group segregated into those in the lumen and those mostly against the mucosa, and within these subgroups species feeding on large and small particles and liquids. Even at this moderate scale of specialization 24 species could be expected in the gut whereas Schad, Kuntz, and Wells (1960) found only 10 species of Tachygonetria. Holmes (1973) chose to emphasize the complementary distributions of Atractis dactyluris with Mehdiella uncinata and Tachygonetria dentata found by Petter (1966) in tortoise guts. But of the 14 species of common nematode studied in detail by Petter, A. dactyluris was the only species possibly involved in competitive interactions, and it became dominant only in older tortoises when many other factors could be changing in the gut.

Of the 43 species of helminth in the scaup alimentary canal,

Hair and Holmes (1975:253) concluded that the community is "composed of a chance combination of ecological specialists." Where they claimed to find inierspecfic competition, only a pair of species was involved and the mean overlap between them was only 14 per cent, but in some cases large numbers of each species coexisted over 20 to 30 per cent of the intestine, suggesting that competition was not the mechanism leading to the small mean overlap value. In one gut Hymenolepis abortiva was abundant but completely overlapped by H. spinocirrosa. For the 13 parasite species in one scaup gut (data for which were given in detail), there was a significant correlation between population size and the number of segments of the gut occupied ($r^2=0.26$, $p<0.10$) indicating niche expansion under population pressure unimpeded by other species. The abundant species were not competitive dominants as there was no sign of complementary abundances in the 10 scaup guts examined, even between H. abortiva and H. spinocirrosa (which showed a positive but nonsignificant correlation in abundance). Hair (1975) concluded that the majority of helminth species in the scaup responded individualistically to gradients in the gut, which I interpret as indicating a lack of positive or negative relationships between parasite species.

Chubb (1964) noted no interactions between Echino-rhynchus clavula and any other parasitic helminths in the guts of fish. Thomas (1964) also concluded after a very extensive study of brown trout parasites that competition between parasites occurs only rarely. In pairwise partial correlations between six helminths over four years and twice a year (165 pairwise comparisons), he found many more positive than negative significant correlations. In a similar set of correlations involving only the four helminths in the alimentary canal whose distributions in the gut largely overlapped, but with up to three seasonal samples per year (264 pairwise comparison) he found only 14 cases that were significantly and negatively correlated (whereas 13 such cases should be expected by chance) and 12 cases showed significant and positive correlations.

Although the three intestinal trematods of the mole show

some differences in distribution along the gut, the probability of colonization for each species was so low that they hardly ever interacted (Frankland, 1959). Less than one per cent of the moles studied had more than one species of the three, and Frankland noted that the distribution of each species, even within an area where it occurs, is very patchy and local. He also found that the presence of these flukes had no effect on the other intestinal helminth parasites of the host.

INTERACTIVE COEXISTENCE

Interactive niche occupation is defined as the substantial shift by one species from fundamental to realized niche in the presence of another species. The cases in 11.1 involving interactive coexistence concern relatively few species: three studies on pairs of species (Cross, 1934: Holmes, 1961, 1962a; Chappell, 1969) and one study reporting co-existence of three species of grill parasite but their exclusion by a fourth (Paperna, 1964). Holmes (1973) provides other example, particularly of competitive exclusion, but the examples in the table are sufficient to illustrate the types of interactive coexistence to be found and the questions raised by such studies.

Three of the four studies showing interactive coexistence involve acanthocephalans (studies 8, 20, 21). In these parasites all nutrients are absorbed through the integument, as in tapeworms, and carbohydrate shortage can be severely limiting (Noble and Noble, 1976). Therefore, these species may be particularly susceptible to exploitative competition for soluble nutrients.

The cases studied by Chappell 1969) and Holmes (1961, 1962a) are particularly convincing examples of competition. In the former case the fundamental niches of the two parasites in stickleback showed a percentage similarity (PS) of 46 per cent, but when occurring together their realized niches had PS=17 per cent. In the rat, the two parasites had fundamental niches that were 64 per cent similar, but in concurrent infections the realized niches were only 10 per cent similar (Holmes, 1961, 1962a).

The excellent experimental studies of Holmes also raise some important questions. The same pair of parasites in rat showed strong niche shifts, particularly by Hymenolepis diminuta; but in hamster there was no indication of competition (Holmes, 1962b). Proportional similarities of distributions along the gut were 45 per cent in single infestations and 46 per cent in concurrent attacks. Also in rat H. diminuta had the broader niche on the gut-length dimension while in hamster M. dubius had the broader niche. It becomes necessary in order to understand coexistence of parasites and their community organization fully to study the full range of coexisting hosts since parasite species and species pairs behave differently in each host. Also where host tissues are damaged and/or immune responses triggered, infection may be transient and a rapid succession of species may be expected. Niche occupation and community development in time may be extremely dynamic.

The dynamic nature to be expected in some parasite communities is fully realized in the gill parasitic community of carp studied by Paperna (1964). The community is composed of four species of monogenean trematode in the genus Dactylogyrus: D. anchoratus, D. extensus, D. minutus, and D. vastator.

Dactylogyrus extensus and D. anchoratus were the better colonizers because they could be infective throughout the year. Thus if young carp hatched early in the year when water temperatures were cool, they were rapidly colonized by these species, and significant numbers began to build up on the gill surfaces (Figure 11.1). However, as water temperature increased, D. vastator became infective and colonization was much more extensive. This species caused gill damage and rapidly made conditions unsuitable for the early colonizers, and they were usually pushed to extinction on the majority of hosts (see Figure 11.1; 50 days after hatching the maximum number of D. extensus found on carp was one). They survived only 10 to 20 days in the presence of D. vastator.

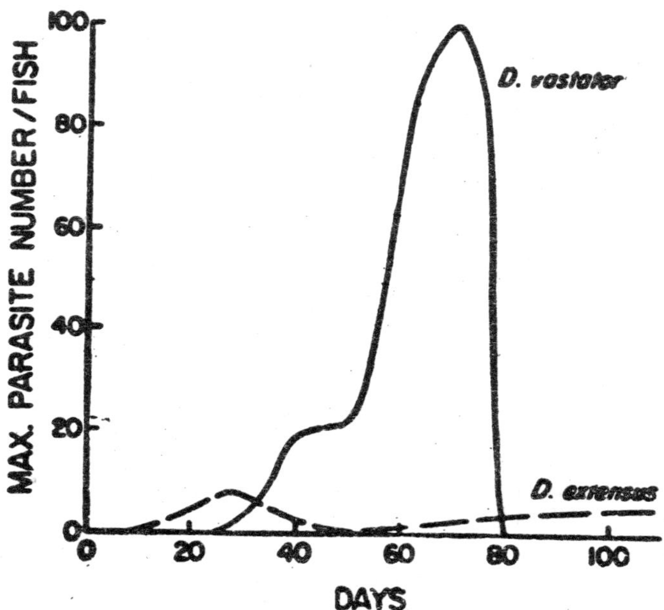

Fig. 11.1. Maximum number of Dactylogyrus extensus and D. vastator found per fish in relation to the number of days after fish hatch.

As the gills became seriously damaged by D. vastator, conditions on the host became unsuitable for it, and after a relatively brief tenure of 44 to 45 days per host, the populations became extinct.

The gills gradually healed and again became available for colonization by the parasities. But a specific immune response to D. vastator eventually developed, the species was eliminated permanently from the community, and the gills were available to the other gill parasites. They very dynamic, non-equilibrium state of the populations and community are also illustrated in Figure 11.2, where fish length is closely correlated with fish age.

This example contains the important elements of non-equilibrium coexistence within patches identified by Skellam (1951).

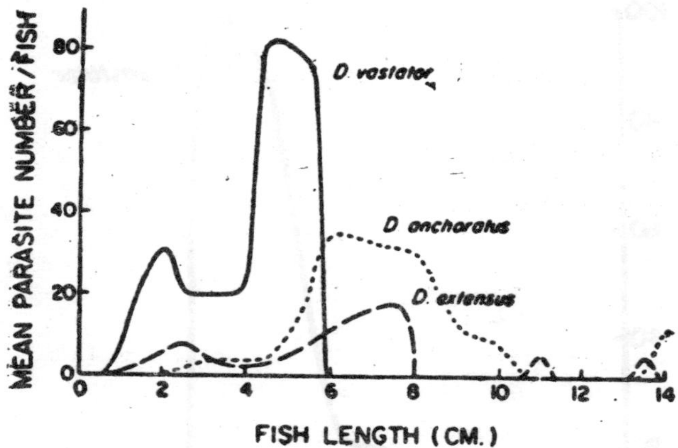

Fig. 11.2: Mean number of parasites per fish for the species Dactylogyrus vastator, D. anchoratus, and D. extensus in relation to fish length.

and Hutchinson (1953) and is similar in some respects to the transient competitive displacement observed by Istock (1967). The competitively inferior species are the better colonizers and appear to have a lower reproductive capacity. New sites are continually becoming available for colonization. Sites rapidly become unsuitable for the competitive dominant, so its competitive edge is repeatedly lost.

Many other examples in the parasitology literature many fit this scenario of transient competitive interactions, but in the absence of good experimental work interpretation of the mechanisms that produce complementary distributions is impossible. For example, Cucullanus heterochrous is evenly distributed in the flounder gut from November to May, but from June to October it is found predominantly in the rectum when C. minutus is present in the intestine (MacKenzie and Gibson, 1970). As the authors point out, we cannot tell if this is a competitive effect or an age effect on C. heterochrous.

Rohde has made a particularly intensive study on species

packing in monogenean ectoparasites of fish. He has repeatedly concluded that many resources remain unexploited, that co-existence is noninteractive, and that "segregation of sympatric species may be due to random selection of niches" (Rohde (1977a: 164, 1977b, c, 1976, 1978). His evidence is convincing. Monogenean flukes usually exist at low population densities (Rohde, 1977a), and although there is a latitudinal gradient of increasing diversity toward the tropics, there is no change in host specificity along this gradient or specificity in site selection on a host (Rohde, 1976). Monogenean species show highly restricted site selection on hosts even when only one species occurs or a host (Rohde, 1977a). When several species coexist, niche overlap is frequently minimal or nonexistent (Rohde, 1977a, b, c). A comparison of niche occupation on temperate versus tropical hosts yielded the conclusion that most micro-habitats utilized by tropical parasites remain empty on fish in cold seas (Rohde, 1976). The mechanisms that result in such narrow niches on a host, even in the absence of competing species, are not fully understood, but the great topographic diversity on fish, particularly on the gill where many monogenean species live, and the considerable complexities of water flow no doubt demand extreme site specificity. This seems to be at least as important as site specificity as a mechanism for increasing population density and the probability of cross-fertilization, an explanation favoured by Rohde (1977a).

Community ecology has been dominated by Gause's principle that species cannot coexist for long if they exploit a very similar set of resources. The evolutionary pressure exerted by competition has been regarded as the major organizing influence in communities (*e.g.*, see Schoener, 1974; Cody, 1974b). But for the majority of parasites listed in Table 11.1 competition appears to be uncommon in the present and to have played an insignificant role in their evolutionary history. Where competition exists, it may frequently result in non-equilibrium transient competitive displacement, a distruptive influence in community development. Based on parasities of plants and animals, fully support these generalizations. For such highly specialized organisms as parasites living in complex environments with steep gradients of resources, even small, randomly generated differences in physiology, behaviour, and morphology may result in optimal exploitation of very different sites in a host or very different hosts. Resources available are so diverse it is importable that a new colonist carries a niche exploitation pattern very similar to a resident species.

12

The Ecology of Desert Plants

The laws of human behaviour are very much in dispute, largely because there are no obvious experimental approaches to them. But animal and plant behaviour can be studied both in nature and in the laboratory, and the science of their ecology should eventually be helpful in the understanding of human relationships, for the basic laws which govern the interrelations among organisms in general must also underlie human behaviour. Ecology is an extremely complex study.

The desert is an ideal area for research. It is usually unspoiled by the encroachment of civilization. Its plant life is sparse enough to be studied conveniently in detail, and it shows clearly and primitively the effects of the physical factors at play in the environment. Most important of all, the desert climate is violent: winds sweep over it unchecked, and its temperature and rainfall swing between wide extremes. Rainfall may vary fivefold from year to year. There are so few rainstorms that the effects of individual rains can be measured. The desert's sharply contrasting conditions can be reproduced in the laboratory for convenient experimental investigation of the germination and growth of plants. And the desert has an unending lure for the botanist; in the spring it is a delightful place.

The most extreme desert in the U.S. is Death Valley. Screened off from the nearest source of water vapour—the Pacific Ocean—by the tall Sierra Nevada, the valley bottom has an average annual rainfall of only 1.35 inches. It has almost no surface water—only a few springs bringing up the scanty runoff from the dry surrounding mountains. Since it is sunk below sea level, Death Valley has no drainage. As a basin which holds and collects all the material that may be washed into it from the mountain canyons, it has accumulated salts in its central part. Seen from above, this salt bed glistens like a lake, but a traveler on foot finds it a dry, rough surface, studded by sharp salt pinnacles which crackle and tinkle as they expand or contract in the heat of the day and the cold of the night.

In the salt plain no green plants can grow: there are only bare rocks, gravel and salt. But on the fringes of the plain plant life begins. Here and there are patches of a lush green shrub—the mesquite. With their tender green leaflets, which suggest plenty of water, the plants seem completely out of place. Actually they do have a considerable source of water, but it is well underground. The mesquite has roots from 30 to 100 feet long, with which it is able to reach and tap underground lenses of fresh water fed by rain percolating down from the mountains.

The mesquite is the only shrub that can reach the water table here with its roots. But a mesquite seedling must send its roots down 30 feet or more through dry sand before it reaches this water. How, then, does it get established? This is one of the unsolved mysteries of the desert. Most of the mesquite shrubs in Death Valley are probably hundreds of years old. Some are all but buried by dunes of sand, piled around them over the years by the winds that sometime blow with great force through the valley. There are places where dozens or hundreds of stems protrude from a dune, all probably the offshoots of a single ancient shrub rooted beneath the dune.

Another Death Valley plant endowed with a remarkable root system is the evergreen creosote bush. It has widereaching roots which can extract water from a large volume of soil. The creosote bush is spread with amazingly even spacing over the desert; this is especially obvious from an airplane. The spacing apparently is due to the fact that the roots of the bush excrete toxic substances which kill any seedlings that start near it. The distance of spacing is correlated with rainfall: the less rainfall, the wider the spacing. This probably means that rain leaches the poisons from the soil so that they do not contaminate as wide an area. We commonly find young creosote bushes along roads in the desert, where the road builders have torn up the old bushes.

During prolonged periods of drought creosote bushes lose their olive-green leaves and retain only small brownish-green leaves. Eventually these also may drop off, and the bush then dies unless rain comes soon afterward. However, it takes a really long drought to kill off all the creosote bushes is an area, They have suffered severely in some areas of the southern California deserts during the drought of the past five years. Because a killing drought tends to remove them wholesale, there are usually only a few age classes of creosote bushes in an area; each group springs up after a drought or during a period of unusual rainfall.

There are other shrubs that master the harsh conditions of the desert, among them the lush green Peucephyllum, which seems to be able to live without water, and the white-leaved desert holly, which grows in fairly salty soil.

Two prime factors control the abundance and distribution of plants: the number of seeds that germinate, and the growing conditions the seedlings encounter while they seek to establish themselves. In the case of the desert shrubs the main controlling factor is the growing conditions rather than germination, for though many seedlings may come forth in a rainy season, few survive long enough to become established. The story is entirely different for the annual plants in the desert.

There are years when the desert floor in Death Valley

blooms with a magic carpet of colour. In the spring of 1939 and again in 1947 the nonsalty portion of the valley was covered with millions of fragrant, golden-yellow desert sunflowers, spotted here and there with white evening primroses and pink desert five-spots. The bursts of flowering are not necessarily correlated with the year's rainfall. For instance, the wettest year in Death Valley was 1941, when 4.2 inches of rain fell, but there was no mass flowering that year or the following spring. If Death Valley is to bloom in the spring, the rain must come at a certain time—during the preceding November or December. There will be a mass display of spring flowers if November or December has a precipitation of well over one inch: in December of 1938 and in November of 1946 the rainfall was 1.4 inches. Rain of this magnitude in August, September, January or February seems ineffective.

Let us consider these annual plants in greater detail. Probably their most remarkable feature is that they are perfectly normal plants, with no special adaptations to withstand drought. Yet they are not found outside the desert areas. The reason lies in the peculiar cautiousness of their seeds. In dry years the seeds lie dormant. This itself is not at all amazing; what is remarkable is that they refuse to germinate even after a rain unless the rainfall is at least half an inch, and preferably an inch or two. Since the upper inch of soil, where all the viable seeds lie, is as wet after a rain of a tenth of an inch as after one or two inches, their discrimination seems hard to explain. How can a completely dormant seed measure the rainfall? That it actually does so can easily be verified in the laboratory. If seed-containing desert soil is spread on pure sand and wet with a rain sprinkler, the seeds will not germinate until the equivalent of one inch of rain has fallen on them. Furthermore, the water must come from above; no germination takes place in a container where water only soaks up from below.

Of course this sounds highly implausible—how can the direction from which the water molecules approrch make any difference to the seed? The answer seems to be that water leaching down through the soil dissolves seed inhabitors. Many seeds have water-soluble germination inhibitors in their covering. They cannot germinate until the inhabitors are removed. This can be done by leaching them in a slow stream of water percolating through the soil, which is what happens during a rainstorm. Water soaking up in the soil from below of course has no leaching action. Some seeds refuse to germinate when the soil contains any appreciable amount of salt. A heavy rain, leaching out the salts, permits them to sprout. Other seeds, including those of many grasses, delay germination for a few days after a rain and then sprout if the soil is still moist—which means that the rain probably was fairly heavy. Still other seeds have inhabitors that can be removed only by the action of bacteria, which requires prolonged moisture. Many seeds preserve their dormancy until they have been wet by a series of rains.

In the washes (dry rivers) of the desert we find a completely different vegetation with different germination requirements. The seeds of many shrubs that grow exclusively in washes (paloverde, ironwood, the smoke tree) have coats so hard that only a strong force can crack them. Seeds of the paloverde can be left in water for a year without a sign of germination; but the embryo grows out within a day if the seed coat is opened mechanically. In nature such seeds are opened by the grinding action of sand and gravel. A few days after a cloudburst has dragged mud and gravel over the bottom of a wash, the bottom is covered with seedlings. It is easy to show that this germination is due to the grinding action of the mud-flow: for instance, seedlings of the smoke tree spring up not under the parent shrub itself but about 150 to 300 feet downstream. That seems to be the critical distance: seeds deposited closer to the shrub have not been ground enough to open, and those farther downstream have been pulverized. Smoke-tree seedlings form about three leaves, then stop their above-ground growth until their roots have penetrated deep enough to provide an adequate supply of moisture for the plant. Thereafter the roots grow about five times as fast as the shoots. Few of these seedlings die of drought, but a flood will destroy most of them; only the oldest and biggest shrubs resist the terrific onslaught of rocks, gravel, sand and mud streaming down the wash.

The ability of the smoke tree to make the most of the available moisture was demonstrated by the following experiment. Cracked smoke-tree seeds were sown on top of an eight-foot-high cylinder containing sand moistened with a nutrient solution. Rain water was then sprinkled on them for a short time. Six seeds germinated, and five of the plants survived and have grown for 18 months in a high temperature with only a single watering midway in that period. Indeed, they have grown better than seedlings which were watered daily!

We have studied the control of germination in great detail in our laboratory at the California Institute of Technology. We have learned, for instance, that two successive rains of three tenths of an inch will cause germination provided they are given not longer than 48 hours apart. Rain in darkness has a different effect from rain during the day. Most amazing is the seeds' specific responses to temperature. When a mixture of rain-treated seeds of various annuals is kept in a warm greenhouse, only the summer-germinating plants sprout; the seeds of the winter annuals remain dormant. When the same seed mixture is kept in a cool place, only the winter annuals germinate. From this it is obvious that the annuals will not germinate unless they can survive temperatures following their germination—and unless there has been enough rain to allow them to complete their life cycle. Since these desert plants cannot depend on "follow-up" rains in nature, they germinate only if they have enough rain beforehand to give them a reasonable chance for survival.

A very small percentage of seeds (less than 1 per cent) germinate after an insufficient rain. Such seedlings almost invariably perish before reaching the flowering stage. On the other hand, more than 50 per cent of all seedlings than have sprouted after a heavy rain survive, flower and set seed. And here we find a remarkable fact: even though the seedlings come up so thickly that there are several thousand per square yard, a majority of them grow to maturity. Though crowded and competing for water, nutrients and light, they do not kill one another off but merely fail to grow to normal size. In one case 3,000 mature plants were found where an estimated 5,000 seedlings had originally germinated. The 3,000 belonged to 10 different species. All had remained small, but each had at least one flower and produced at least one seed. This phenomenon is not peculiar to desert plants. In fields of wheat, rice and sugar cane, at spots where seeds happen to have bcen sown too thickly, all the seedlings grow up together; they may be spindly but they do not die. It is true that in gardens weeds often crowd out some of the desirable plants, but usually this happens only because these plants have been sown or planted out of season or in the wrong climate. Under those conditions they cannot complete with the plants fully adapted to the local growing conditions—plants which we usually call weeds.

We must conclude, then, that all we have read about the ruthless struggle for existence and the "survival of the fittest" in nature is not necessarily true. Among many plants, especially annuals, there is no struggle between individuals for precedence or survival. Once an annual has germinated, it matures and fulfills its destiny of forming new seed. In other words, after successful germination annual plants are less subject to the process of "natural selection." Very likely this accounts for the fact that so few of the desert annuals seem to show adaptations to thc desert environment. This does not mean that the plants have avoided evolution, but the evolution has operated on their seeds and methods of germination rather than on the characteristics of the grown plants. Selection on the basis of germination has endowed the plants with a remarkable variety of mechanisms for germinating, and at the same time it has made them show to germinate except under conditions insuring their later survival. The opposite is true of the cultivated plants that man has developed: has selection has favoured the plants that germinate most easily and quickly. This has given us the wrong perspective on the significance of germination in plant survival.

We return now to our original theme: Can the ecology of

plants in the desert teach us anything about human ecology or human relations? At least one moral stands out. In the desert, where want and hunger for water are the normal burden of all plants, we find no fierce competition for existence, with the strong crowding out the weak. On the contrary, the available possessions—space, light, water and food—are shared and shared alike by all. If there is not enough for all to grow tall and strong, then all remain smaller. This factual picture is very different from the time-honored notion that nature's way is cut-throat competition among individuals.

Actually competition or warfare as the human species has developed it is rare in nature. Seldom do we find war between groups of individuals of the same species. There are predators, but almost always they prey on a different species; they do not practice cannibalism. The strangler fig in the tropical jungle, which kills other trees to reach the light, is a rare type [see "Strangler Trees," by Theodosius Dobzhansky and Joao Murca-Pires; Scientific American, January, 1954]. Even in the dense forest there is little killing of the small and weak. The forest giants among the trees do not kill the small fry under them. They hold back their development, and they prevent further germination. In a mountain forest in Java it was observed that the small trees living in the shade of the forest giants had not grown after 40 years, but they were still alive.

Hundreds of different species of trees, large and small, grow in a tropical jungle. This diversity of vegetation is one of the jungle's most typical characteristics. Some trees grow faster, taller or wider than others, but these growing characteristics, which we have always considered as useful adaptations in the struggle for existence, do not really control the trees' survival. If they did, we would find very few species of trees in a jungle, and there would be an evolutionary tendency for these trees to become taller and taller. Actually the tallest trees are found not in jungles but in more open forests in temperate climates; remarkably enough, tropical jungles often have no particularly high or . ~ge trees. All this shows that selection does not work on the basis of growth potential. It works on the ability of plants to grow and survive with very little light.

In our minds the struggle for existence is usually associated with a ruthless extermination of the less well adapted by those better adapted—a sort of continuous cold war. There is no cold war or even aggression in the desert or jungle. Most plants are not equipped with mechanisms to combat others. All plants grow up together and share whatever light or water or nutrients are available. It is only when the supply of one of these factors becomes critical that competition starts. But it appears likely that in the jungle, as in the desert. survival is taken care of by the control of germination. Competition and selection occur during germination, and we can speak of germination control of the plant community—comparable to birth control in human society.

Apparently evolution has already eliminated most of the plant types that are unable to compete successfully. Fast-growing, show-growing or tall plants all have the same chances once they have germinated. The struggle for existence is not waged among the well-established plant forms but tends to eliminate new types which germinate at inopportune times, have a decreased ability to photosynthesize or are less frost-resistant. This explains why so few plants die in the desert from drought or in the jungle from lack of light or in cold climates from frost.

As a general moral we conclude that war as man wages it finds no counterpart in nature, and it has no justification on the basis of evolution or natural selection. If we went to describe the process of control of the plant population in human terms, we should talk about birth control.

13

Forest

Introduction

Of all the environmental problems facing the country, the problem of deforestation has received the maximum public attention. Ironically, no set of government policies in the field of environment has attracted greater public criticism than the policies for afforestation.

Social forestry programmes have been launched by several states to promote afforestation on essentially non-forest lands, that is, on private farms and on village commons. Several state governments have stepped up their own afforestation efforts and are also planning (in most cases, without much public discussion) to hand over large tracts of degraded government forest lands to industrial firms for afforestation.

All these programmes have one thing in common the planting of trees to meet the needs of urban and industrial markets while the glaring fuel and fodder crises facing the common person continue to grow. The species chosen-like eucalyptus, teak and pine-also do little for ecological restoration, for enhancing soil fertiliy or for soil and water conservation.

The very same process of commercialising the forest

resources base that has led to widespread deforestation in the country is today the motive force behind the government's afforestation programmes being carried out with the full support of foreign aid agencies, India's afforestation is, thus, as anti-people as is its deforestation. Rural women, for instance, whose lives revolve around the collection of fuel and fodder, have almost nowhere been involved in these programmes.

People are responding with greater alacrity to voluntary afforestation efforts aimed primarily of meeting people's needs of fuel, fodder and small timber—and popular protests grow against both government sponsored deforestation and afforestation. Few in the government have the imagination to realise that afforestation of India's degraded public lands could form the core of India's largest reforms, anti-poverty and employment generation programmes.

Social Forestry

Social forestry is a term used by the National Commission on Agriculture in 1976 to denote tree-raising programmes to supply firewood, fodder, small timber and minor forest produce to rural populations. Nearly a decade later, it is emerging as the most controversial initiative of the Indian government.

Critics of the programme strongly contend that the wood produced from social forestry programmes is ending up in urban and industrial India instead of the poor in rural India, reducing rural employment and land under food production, and promoting absentee landlordism.

The Union Government and the World Bank are discussing a $600 million proposal for a nationwide social forestry programme which will cover 13 states including a number of those in the Northeast.

Ambitious social forestry programmes have already been launched by several state governments, most of which have been financed by foreign aid agencies like the World Bank, US

Agency for International Development (USAID), Canadian International Development Agency (CIDA) and the Swedish International Development Authority (SIDA). Forest departments in most states have set up separate social forestry wings.

Growing Criticism

Despite these achievements, present and projected, doubts and criticism about the actual benefits of social forestry have grown into a shrill ehorus.

Social forestry programmes have mainly three components: farm forestry, encouraging farmers to plant trees on their own farms by distributing free or subsidised seedings; woodlots planted by the forest departments for the needs of the community, essentially along roadside, canal banks and other such public lands; and community woodlots planted by the communities themselves on community lands, to be share equally by them.

Judging by the World Bank's own mid-term reviews of social forestry projects in Uttar Pradesh and Gujarat, big farmers are emerging as the primary beneficiaries. In Uttar Pradesh the World Bank assisted social forestry programme overshot its farm forestry target by 3,430 per cent a phenomenal and unprecedented degree of success for a government programme. However, efforts in the creation of community self-help woodlots achieved only 11 per cent of the target, a disastrous record even though this component of social forestry programmes is aimed most at meeting the needs of the neediest in the rural communities, the landless and small and marginal farmers. In Gujarat, too many forestry targets were exceeded by more than 200 per cent whereas community self-help woodlots met only 43 per cent of the target.

In communist-ruled West Bengal, the best state according to the World Bank, the creation of village woodlots by the forest department has fared well, achieving 109 per cent of the target. But these are not self-help community woodlots. Landless families have been given small plots of government wastelands for tree planting. The Marxist lead United Front government has also ensured involvement of its party cadres in the Programme, thus evoking a good grassroot response. Even then farm forestry in West Bengal has fared better, achieving 200 per cent of its target.

The Agriculture Ministry itself admits that the percentage of seedlings reaching farmers with landholdings of more than 2 hectares has been 61.5 per cent in Haryana and 57.6 per cent in Gujarat. Jammu and Kashmir and West Bengal have fared better: this percentage there is down to 35.3 per cent and 19.5 per cent respectively.

A significant result of the success of farm forestry the "spontaneous response of small farmers or communities to the commercial incentive of rising prises (for wood) and the perception that 'farming fastgrowing short rotation trees as a cash crop can be as profitable as farming some of the traditional cash crops", as a World Bank document puts it is the conversion of large tracts of prime agricultural land to tree farming. In Uttar Pradesh, the average area to be planted annually under farm forestry was targeted at 2,800 hectares over a four-year period between 1979 and 1983, but over 98,000 hectares have actually been planted. Seedlings sold by the social forestry wing have been cornered by farmers owning over 4 hectares. In eastern Uttar Pradesh where landholdings are extremely fragmented, peasants extremely poor, and the energy crisis extremely acute. farm forestry has fared poorly. This is yet another indication that it benefits better off farmers and does little to meet the energy needs of the poor.

The fact that seedlings are distributed free or are heavily subsidised essentially means that the programmes, started with the professed aim of meeting the dire fuel and fodder needs of the poor, are becoming a scheme of subsidies to support lucrative cash cropping by the rich. Additionally the fact that seedings are often not available immediately after the rains means that small farmers with unirrigated lands cannot take as much advantage of subsidised seedlings as rich farmers with irrigated lands can. In many states, irrigated lands have come under tree farming and in at least one state there are now demands to can free farming on irrigated lands so as to affect food production seriously.

Farm Forestry

A major reason for the distortions in social forestry programmes is the acute shortage of building poles in the urban market and of pulpwood, required by the rayon and paper industries. The active propagation of farm forestry by the government and the financial incentives it holds out has come as a windfall for the big farmer. Labour needs go down when farmers switch from short-term agriculture to long-term tree farming and they can even contemplate becoming absentee landlords. If they plant a tree like eucalyptus which is not palatable to animals, supervisory problems are also reduced.

However, for the landless in the village this can be a real disater. For instance, in Punjab an influential farmer with over 100 hectares of land switched over form cotton to eucalyptus farming. The change in the type of biomass being produced created immediate energy problems for the farmer's labourers who earlier got cotton sticks regularly for fuel. Now they get nothing. After five to seven years, when the eucalyptus is harvested, they might get the lops and tops, as the trunk will go to the urban market and that only if the farmer is unable to sell the lops and tops of the tree. There is every reason to believe that with social forestry. India is getting into a new green revolution which will probably increase wood in the villages but will also make the poor even poorer in terms of energy.

A major reason why farm forestry has become such an overwhelningly dominant component of social forestry is what the World Bank calls, "Project economic justification". Government strip plantations cost about Rs. 4,200 per hectare and village woodlots Rs. 3,000 per hectare and the internal rates of return for such plantation range from 12 per cent to 15 per cent. By contrast, the cost of farm forestry is about Rs. 1,600 per hectare and the internal rates of return are of the order of 25 per cent to 30 per cent. Concludes a World Bank document: "The comparison clearly reinforces the conclusions now being drawn by the community forestry wings project management that farm forestry is likely to be the least costly and economically the most effective approach to afforestation of the rural areas and that a major thrust of project development in the future should be in this direction."

One of the best indicators of the insensitivity of the World Bank to the needs of the poor is the fact that all its documents harp on the importance of fuelwood almost without any mention of the fodder problem, while there is no reason to believe that the fodder crises is any less severe than the fuel crisis, if not more. Even the weeds that take over degraded lands can provide some fuel but they provide no fodder. They spread only because they are not palatable to animals. The World Bank and foresters are happy to spread eucalyptus, a plant which specifically cannot provide fodder—a plant which socially speaking has all the characteristics of a weed—and yet make loud claims about the social relevance and success or social forestry.

The champions of "commercialised social forestry" assert that the gains of farm forestry far outweigh the losses. By increasing the wood available for urban and industrial purposes pressure on natural forests will be reduced. Pro-eucalyptus economists argue that once wood supply overtakes industrial and urban demand, prices of wood will fall and farmers will diversify their plantations. But they do not answer why farmers will go on to produce wood for the poor who cannot pay.

That the commercial wood market is of key interest to the World Bank can be seen form the following paragraph from one of its own documents: "Marketing studies are seen as a critical issue in social forestry projects being funded by the Bank in India. It is becoming increasingly obvious that the rising prices for poles and fuel wood have been one of the main factors in stimulating farmers' interest in tree planting. Early definition of the size and geographic location of such cash markets can influence the siting of tree nurse ies and the deisign of appropriate extension programmes."

Revealing Survey

A model study prepared for the World Bank by the International Institute for Environment and Development on the wood market in Gujarat claims that World Bank officials are worried that if existing cash markets were to be satisfied in the conceivably foreseeable future, this markets saturation could have an adverse effect on farmer and community response to farm forestry. The study, therefore, looks into the existing and potential market structure for future project planning. It concentrates entirely on the commercial wood demand, urban, firewood demand, demand for construction poles, sawlogs, wood and pulp, and paper, and other wood products like activated charcoal, catechu (a tanning extract) and 'katha' (a condiment used in the preparation of betal nut). But no where does the study talk about the non-commercial firewood demand of the poor, as if that was never an objective of the social forestry project in Gujarat.

The study, which is to be used by the World Bank as a model for planning social forestry programmes in other states, goes into great detal about how the commercial wood market can be expanded to continue to promote social forestry.

"In addition, enterprising and innovative individuals are actively considering potential markets that could be tapped, should any surpluses become apparent. The demand for wood products in India is exceptionally high and the lack of a sufficient raw material base has greatly hindered market development. If the raw materials become available through social forestry activities, it can be expected that new forest product industries will develop, particularly pulp and paper, wood-based panels, and saw milling industries.

"Because a relatively large amount of wood will become available over a short period of time, there is some concern that the current marketing structure may fail to protect vulnerable small farmers with little experience. A promising sign is that wood growers' cooperatives have been formed in a few areas (emphasis in original)".

The study also talks of various new opportunities being discussed in Gujarat to expand the commercial wood market. For instance, it reports that there has been some discussion among farmers, particularly in Bhavnagar District, on establishing small-scale village mechanical pulping devices that would be capable of handling eucalyptus. They have also been discussing with the State Agro-Industry Board and the Gujarat Energy Development. Authority the idea of wood-based electricity generating power station whose power would be purchased by the State and fed into the central grid. Jyoti Ltd. has been working on such a dendro-thermal power plant. This company has also designed a wood gasilier-powered irrigation pump, which will extend the wood marked further. The economics to the Jyoti system appears to be quite favourable, depending on the availability of wood. Units of varying sizes are expected to be put on the market in the near future. Another company has approached the forest department for establishing a hardboard plant in Bhavnagar area. Good quality eucalyptus timber could also be used by the ship building industry along the coast. Essentially this ahds up to a wood production, marketing and development scheme.

Admission of Failure

Access to fodder, fuelwood and building wood essentially means access to land and social forestry projects. If they are really to serve the neediest, must be directed to landless, small and marginal farmer households. The World Bank admits in its latest global review of World Bank financed forestry activity that it has failed in this respect; "Ensuring adequate supplies of fnelwood, building poles and fodder for the landless, who in some developing countries number more than 25 per cent of the population, is probably the most difficult forest policy issue that most developing country governments face. Various approaches have been tried, such as governments-sponsored establishment of road-side plantations or woodlots with preference in allocation of employment and output to landless families. The problem of creating viable employment for the landless is an issue which can only be resolved by general economic growth, by movement of people out of the rural areas into industrial employment, by resolution of inequities in land distribution. by growth of the service sector, by development of markets for processed primary products and by other development-oriented interventions. Forestry's role in this area will remain marginal. A second constraint is that of how to provide effective incentives to persuade small farmers to take up tree planting is to create a supply of wood for domestic consumption."

The landless, in fact, are distrustful of social forestry programmes. The World Bank itself lists a number of reasons whose response to community forestry efforts has been poor: heterogeneous village communities; mistrust of the system for ensuring equitable distribution of the produce from community woodlots, disputes amongst farmers concerning availability of the village common land for the establishment of village woodlots, and shortage of panchayat funds for payment to labour and protection of plantations. "The 'commons' problems is particularly difficult issues as it affects the willingness of rural communities to cooperate in the establishment of village woodlots, and to protect them from grazing animals", admits the Word Bank.

Surveys in Tamil Nadu show that the cattle-owning landless households are against woodlots on village commons because they are afraid this will further reduce the availability of grazing lands, block the ancient right of passage for carts and reduce the supply of brushwood (the poor person's fuel). Also experience has shown the poor that the produce from such community efforts seldom reach them. Both Madhya Pradesh and Uttar Pradesh have tried to ensure equitable distribution of the usefruct by drawing up elaborate contracts between the panchayats and the social forestry directorate. But as experience in a village in Sultanpur District shows despite a list of the "poorest of the poor" being made by the village pradhan the actual distribution of firewood was in favour of the landed: the landless got nothing. Uttar Pradesh and its donor World Bank, are not even pursuing such contracts in other villages because they consider it a futile exercise.

The World Bank's reaction to this failure exemplifies its limited commitment to get social forestry programmes serving community needs. The Gujarat social forestry appraisal report describes how project staff are resorting to management decisions to "increase the size of the harm forestry programme, to reduce expectations under the village self-help component and to increase the forestry department, supervised village woodlot componet". In other words, if self-help woodlots have not met with success, forget the poor and switch to farm foresty. J. Bandyopadhyay, one of the authors of the Kolar study, strongly questions the propriety of international borrowing for financing social forestry programmes mainly executed by rich farmers, and whose benefits do not reach the poor. If social forestry is so profitable why should there be any subsidies, especially those which make the country indebted to foreign bodies, he asks.

In many states additional resources for social forestry have been mobilised by linking if to the National Rural Employment Programme (NREP) and the Integrated Rural Development Programme (IRDP). In Delhi, Madhya Pradesh, Uttar Pradesh and Gujarat, trees are being planted on public lands with seedling provided by the forest department and the wages component funded under NREP. In Gujarat the self help component of the community woodlots scheme picked up momentum after the forest department decided that the poor and the landless should not be expected to volunteer their labour but should be paid out to NREP funds. However. this has only led to some token social forestry with an eye to obtaining more funds. In Pudukottai District of Tamil Nadu, for instance, it was found that because of the stipulation for receiving grants under IRDP or NREP, most of the panchayat unions had reluctantly raised plantations, but these were neglected,

The World Bank and the forest departments do not admit that tree farming is eroding rural employment but they do not produce any hard data or studies either to prove otherwise. Figures produced show that social forestry is creating employment but no comparative studies have been published to show whether his employment is less or more than cultivation would generate in irrigated and non-irrigated conditions. The Kolar study estimates that for each hectare of land shifted from food crops to eucalyptus, there is a loss of about 250 person days of employment per year. Other experts, however, argue that forest plantations can increase employment for the rural landless when compared to food crops.

A major failing of the social forestry projects is the lack of involvement of poor women who ought to be the main beneficiaries. It is women who collect fuel and fodder every day for the family and the male orientation of the projects has, not surprisingly, turned them into cash-generating rather than basic needs generating exercises.

Shifting Priorities

Whatever propaganda the World Bank officials may continue to use for their ignorant admirers in Washington, in India they have kept a low profile. In private, World Bank Officials admit that, having learnt their lesson, they are now stressing afforestation of degraded forest lands as an important component in social forestry programmes to meet primarily the fuel needs of the poor. They insist that the World Bank is now increasingly emphasising individualisation or privatisation of trees grown on degraded forest lands. Right to the usufruct of the planted trees will be given to poor households living near such lands.

In West Bengal, for instance, small plots of government-owned wastelands are being allocated to landless families for tree planting, the landless do not get the title to the land, but they are given full right to the usufruct of the trees they plant, Both wages and seedlings are given at the planting stage. Stage States like Gujarat and Maharashtra have also developed schemes to involve the landless in tree planting but it is too early to evaluate these schemes. Certain reports indicate that the forest departments show little interest in these schemes and tribal families are reluctant to take advantage of them for fear of getting involved in something that may only harm them in the long run. Voluntary agencies may have to play an important role in organising the poor to take full advantage of these schemes.

In any case only those landless families can be involved in the schemes which live close to the degraded forest lands. Moreover, any major move to settle the landless on degraded forest and will face strong opposition from wood-based industries which are pushing for allotment of such lands to them, to meet their own wood requirements.

A fundamental reason for the failure of the social forestry programme to help the most needy is that it expects an unrealistic degree of collaboration between the villagers who use the forests and the forest department personnel who police them. Many villagers in Gujarat, Madhya Pradesh, Uttar Pradesh and elsewhere are reluctant to let the forest department afforest their common lands for fear that they may claim the afforested area as theirs. Poor rural families have even been reluctant to accept saplings distributed free for homestead plantations through social forestry projects for fear that planting them will somehow give the forest department a claim to the land.

Foresters too, do not relish the new role thrust upon them of wooing the rural public as partners in afforestation. A young divisional forest officer in Madhya Pradesh sums up: "We used to be kings (of the forests), now we have to go as beggars (to persuade people to plant trees)." It is clear that there has to be a fundamental change in objectives and attitudes if the social forestry programme is to make a lasting impact in areas where such an impact is of prime importance.

Government Forest Lands

Attention in the last two years has mostly focussed on social forestry programmes, programmes mainly promoted on private farmers and panchayat lands. Little public attention has focussed on the management of 75 million hectares of forest lands under the control of the forest department. Inside government committees, the paper industry, facing an unprecedented raw materials crisis is arguing for control over large tracts of degraded forest lands which it claims it could afforest and generate its own raw materials. Until now the industry has only dipped into the vast natural forests, in most cases at throwaway prices, and caused massive deforestation, often in collusion with forest officials and money-hungry politicians.

Paper is a vital commodity. The per capita consumption of paper—often regarded as an index of literacy and standard of living—is around 2 kg per person per year, a pitiful amount when compared to the 268 kg in the US, 124 kg in the UK and 115 kg in France and even 11 kg in Thailand and 10 kg in Egypt. The paper industry in India has steadily increased its capacity from 0.14 million tonnes in 1951 to 2.16 million tonnes at the beginning of 1983. These were then a total of 179 paper miles. About 20 per cent of the present paper-making capacity was in existence before the 1950s. Another 50 per cent came into existence between 1950 and 1970. The remaining 30 per cent was added during the 1970s. The period was marked by a drastic reduction in the growth rate compared to the 1950-70 period. The capacity added during the 1970s mainly came through schemes involving addition of balancing equipments in large mills and through the proliferation of small mills based on agricultural residues.

The Indian paper industry, like the Chinese, is unique for its wide range of mill sizes. Mill capacities range from 5 tonnes to 10 tonnes per day in the smallest units to about 250 tonnes per day in the largest mill. The latest newsprint mills are of 300 tonnes per day capacity. But even though there are many small paper mills in the country, about 65 per cent of the total installed capacity is represented by the large integrated mills with plant sizes of over 20,000 tonnes per year, which are only 16 per cent of the total number of mills in the country.

The growth of large mills was greatly inhibited during the 1970s because of steeply escalating energy and investment costs and even more because of growing shortages of raw materials like bomboo. A "redeeming feature" of the paper industry during the 1970s, according to the Development Council of the Pulp and Paper and Allied Industry (DCPPAI), constituted by the government, was the emergence of many small paper mills with an annual capacity of less than 10,000 tonnes based on unconventional materials like agricultural resides (rise and wheat straw etc.), cotton rags, hemp, jute cuttings, and grass. Most capacities created in the past decade are based on agricultural residues, helped by tax concessions offered by the government for use of unconventional raw materials. Small paper mills use on an average 55 per cent agricultural residues 40 per cent waste paper, and 5 per cent purchased pulp and other long fibre materials like cotton linters and rags.

Waste paper recycling is also increasing in India, even though it is still low (only 20 per cent) as compared to many European countries and Japan where it is as high as 50 per cent. Raw material and energy shortages, particularly electricity: have also become major constraints in existing paper mills based on bamboo. Capacity utilisation in the paper industry has steadily declined from a peak of 98.9 per cent in 1970 to 68 per cent in 1981. According to the Indian Paper Makers Association (IPMA), capacity utilisation in 1983 was only 58.1 per cent. Thus even with this large capacity, the paper and paperboard industry could achieve a production of only 1.11 million tonnes in 1983, valued at Rs. 1,200 crore. To achieve this production the mills used about 3.09 million tonnes of forest raw materials.

Large paper and paperboard mills are fully integrated to pulp production which is now based mostly on bamboo and hardwoods. These mills are generally located close to forests. The general policy is to be self reliant in fibre supply and use market pulp only as a "buffer" material. Bamboo, essentially a medium fibre material, was then available in the country in abundance. But now with bamboo resources shrinking large mills have no other option but to mix hardwoods with bamboo, even though there is a decline in pulp quality. Use of soft woods is still very limited. The 28 large integrated mills, with an installed capacity of 1.4 million tonnes, use on an average 60 to 65 per cent bamboo, 30 to 35 per cent hardwoods, and very minimally other fibrous materials. The paper industry claims that it today accounts for about 2 per cent of the country's annual consumption of wood and 51 per cent of bamboo.

Bamboo Crisis

The growing bamboo shortage poses production and ecological problems: over exploitation of existing forests and increased exploitation of remote forests.

According to the Pre-Investment Survey of Forest Resources, annual production of bamboo varies from 2.5 million tonnes to 3.0 million tonnes. About 1.6 million tonnes is supplied to existing paper and pulp mills and the remaining is used for construction by poor rural and urban people. Bamboo is popularly known as the poor person's timber and is already in short supply because of the paper mills. No sizeable bamboo resources remain to sustain new paper and pulp mills or even to feed the expansion of the existing ones.

Forests in Danger

In such a raw material crisis, all companies, private and public sector, want to head towards the last forest frontiers like Bastar, the Northest and the Andamans. The government expects that the three paper mills of the public sector Hindustan Paper Corporation at Tuli in Nagaland and at Nowgong and

Cachar in Assam will soon increase the country's paper production capacity by 233,000 tonnes of paper and paper-board. These plants will use bamboo as their main raw material. Lack of transport infrastructure is the main stumbling block in the exploitation of the 'green mines' of the Northeast but, once better roads are built this area is certain to be devastated. Meanwhile, a public sector paper mill is being considered by the Andhra Pradesh Industrial Development Corporation for Kakinada. This paper mill will use wood chips from the virgin Andaman Islands as its main raw material.

State governments in their bid to attract industry and politicians in their bid to make big money have often given away bamboo forests on lease at throwaway prices. The Orient Paper Mills of Ambai was initially given bamboo at dirt cheap prices, 37 paise per tonne by the Madhya Pradesh government. The Seshasayee Paper Mills in Tamil Nadu was paying Rs. 6.89 per tonne until 1970, Rs. 11 from 1870 to 1975, and Rs. 22 plus 5 per cent for administrative expenses since 1975. Paper mills in Karnataka were paying Rs. 15 per tonne for bamboo when the poor could purchase it in the open market only at Rs. 1,200 per tonne.

In recent years, however, many state governments have begun to revise royalty rates steeply upwards. Inevitably all kinds of cheap raw materials. When the Orient Paper Mills closed down its Brajrajnagar plant in Orissa, ostensibly because of the workers' agitation for more bonus, it was widely rumoured that the management had done so deliberately to bring in cheaper bamboo from Orissa and pressurise the Madhya Pradesh government which had suddenly jacked up bamboo rates to Rs. 102 per tonne in 1981. There is also the case of a West Bengal paper mill which had approached Bihar government for allocating forest areas for its raw material requirements after its own sources in West Bengal ran short. But when the Bihar government refused to do so unless the mill was set up in Bihar, all that the company had to do was to send its truck to the Bihar-West Bengal border.

Thousand of headloaders would illegally deliver wood to it every day.

In 1980, the Public Accounts Committee of the Karnataka Legislature had studied the sale of eucalyptus plantations to the Gwalior Rayon Silk Mills—the rayon industry is another major user of pulp at Rs. 24 at tonne which was lower than the cost of production of Rs. 44 a tonne. The Committee suggested that the price was very low and in February 1981 the price of eucalyptus was fixed at Rs. 81 a tonne. But five days later the then Chief Minister R. Gundu Rao, gave oral instructions that the forest department should continue to sell forest produce at the old rates to industries after securing a bank guarantee representing the difference between the old and new rates. In April 1931 fresh instructions were issued to continue supplies at old rates against a bank guarantee of only Rs. 5 lakh. The Committee indicated the Chief Minister for saying the operation of the new rates, declaring it illegal and staying that it had led to a loss of Rs. 22 crore. The Committee also expressed dismay over the disappearance of the file in which Gundu Rao's oral instructions, staying the revised rates, were recorded.

But many state governments have not learnt the lesson in charging low royalties in the first place. The new 80,000 tonne per year newsprint plant of the public sector. Hindustan Paper Corporation, will pay the state of Kerala only Rs. 11 per tonne of green wood of eucalyptus, far below the cost of plantation. The Kerala government will supply eucalyptus and reeds on a sustained basis in adequate quantities. The management, however, does not rule out acute raw material shortages unless a massive plantation programme is taken up. Part of the mill's requirements are already being met through imported pulp.

Importance of the Forest

Forest is needed to conserve soil, water, air, to maintain the required level of oxygen, carbon dioxide, etc., to resist pollution of air, water and soil; to maintain nature's equilibrium between living things; to supply medicines; to have biotic check on pest and diseases; to screen irritating smoke, dust, moving sand, etc., to absorb sound; to cut off gloves and divide lands for directing traffic; to plant barren hills to check erosion; to ensure privacy by planting small species of trees; to act as store house of food; to check floods and droughts; to induce rain; to increase further agricultural yields; to provide organic chemicals, employment, income and profits; to conserve gene-pools and over all for holistic management of the ecosystems. It is the base camp of life's journey to the planet. Yet, the allocation of funds in Central and State budgets for forest is a pittance compared to the allocation to irrigation, energy and other sectors.

Depletion of the Forest

The depletion of the forest is due to its exhaustible use as fuel wood without consideration of its growth; its use for industrialisation, towns, living habitat for exploding population, refugee settlements: commercial exploitation for housing, household goods, agricultural uses, construction of roads, railways, airports, dams, mining; destruction by army and illegal felling due to corrupt officials and politicians, etc. Vaumahotsva, Social forestry, reclamation of land, Chipko and Appiko movements, forest week, etc., are nothing but apology of the mankind to the nature for the age long holocaust done by him on the fragile forest ecosystems.

Problem of Forestry in Few Indian States

To draw some of the fundamental conclusions regarding the forest problems of India it will be better to discuss some of the appealing problems of the forest of the states and their efforts in the field of conservation.

Nagaland

The deforestation of Nagaland is taking place at a frightening pace. Nagaland government report of 1983 shows that total forest area were 91,525 ha of which 29,255 ha were reserved and rest were protected. Private forests covered 1,81,150 ha of land. The State has 10 vaneer mills, 100 large and medium industries, 200 small and 50 mini saw mills with the annual deforestation of 18,000 ha. This way, Nagaland would run out its exploitable forests within 10 years. The only forests which will survive are, those which cannot be exploited because of difficult terrains or inadequacy of communication. Besides, the smuggling, here instead of felling trees for stretches, forests are thinned out and which can on every hectare of land.

Himachal Pradesh

About 50,000 trees mostly spruce and fir, measuring 1,80,000 cu m are cut down every year to meet the requirements of the orchardists in the State. It is an apple state and due to the pressure from apple lobby it is very difficult for the Chief Minister to take any stringent decisions. But due to pressure from Central Government, the Chief Minister has decided to switch over from wooden boxes to cartons for packing fruits especially apples. The cartons are now supplied on subsidised rate and the government has decided to set up one cardboard factory. The popular trees are also planted in captivity to ease the pressure on packing woods.

Madhya Pradesh

In 1986, the Chief Minister of Madhya Pradesh has banned felling of trees in 11 districts which does not include Bastar district. Unofficially, it is reported that more than 500 truckload of forest logs are smuggled out of the district every month with the collusion of the officials. What is more alarming is that not only the dry or dead woods were cut, and even the green trees were not spared by these unscrupulous elements. They exploit the tribals who have free access to protected forest areas and are allowed to carry a headload of fuel wood. This headload is often converted into truckload, and these trucks cross the state border unchecked.

The new onslaught is now on the sal forest of Baster (spread over 22,703 sq km area) setting up the hydel power project or for the Craze opening industry. It is estimated that Bodhghat

Hydel Project will damage about 45 lakh green trees, including at least 17 per cent sal forest. Here is an important fact that sal trees cannot be grown by planting saplings, but can grow naturally when the sal seed falls on forest ground.

Gujarat

Gujarat forest department has drawn a special scheme for forest protection and employment for the tribal population. Although this is on meagre scale but it can be model for 14 per cent tribal population of the State and also to the other States of the country.

In order to pay the special attention on the development of the 194 forest village colonies of the State, the chief conservator has adopted a new model. The social defence scheme provides employment to the Adivasi families on a permanent basis in forestry. Each year beneficiary family is allotted 1.5 ha land on which they undertake afforestation according to the instructions of the forest department. By this way, they are assured 25 days employment in a month. These beneficiaries are considered as participants of this programme. They will be given 20 per cent of the net profit from tree cutting. This scheme provides them a livelihood means for maintenance. The scheme has benefited 311 families in 1985-86. A scheme for awarding prizes to these individuals and Panchayats helping in nabbing the forests criminals is also in force.

West Bengal

In this State small plots of government owned wasteland are being allotted to landless families for tree planting. The landless do not get the tillership of the land but they are given full right to use trees they plant. They are provided seedings at the planting stage. The same philosophy has been adopted by Maharashtra and Bihar also.

Tamil Nadu

Tamil Nadu shows an interesting report on social forestry. In

Chengalpattu district, a eucalyptus hybrid plantation was begun in 1980. However, immediately after planting, 5,000 saplings were uprooted by villagers. An evaluation of 20 years of farm forestry of the same district (1960-80) in which 69 species were listed for plantation only two, babul and eucalyptus hybrid dominate and the forest department persisted with eucalyptus even when the villagers demanding casuarina because it was more suitable for fuel.

The Mangrove of Different States

India has about 7,000 km² of its area under mangroves forest. These consist of deltaic mangroves along the eastern coast, estaurine mangroves along the western coast and insular mangroves on the Andaman and Nicobar islands. These forests are very important because it has got very low floristic diversity. They are under serious demographic pressure and creating ecological hazards due to their reckless cutting and extinction due to human interference.

Measures for Improvement of Forest Ecology in India

It is very difficult to give some concrete suggestions for improvement of the forest ecology of India. However, few measures can be adopted for maintaining the biological treasure of India.

(a) The Hill development plan should be both ecologically sustainable and economically viable for the rural community. There should be harmonious relationship between man and forest. Natural mixed forests should be allowed to grow and mono-cultural is to be given up. Cultivation of agricultural crops in areas above 15° slopes should be stopped and planted with five "Fs" (food, fodder, fuel, fertilizer and fibre) trees.

(b) Industry producing dust and fumes should not be encouraged in the hilly areas. Household industries like spinning, weaving, knitting, embroidery, carving, watch making, etc., are best suited for the same area.

(c) Polyculture, instead of monoculture can be encouraged. Besides this, cooperation and collaboration of the local people must be sought at each and every level of policy formulation and implementation undertaken to save the areas from landslides and floods.

(d) The National Forest Policy of India has given emphasis to restore ecological balance, optimum utilization of land, water, livestock and human resources. It is having the three priorities *i.e.*, priorities of social forestry for meeting the domestic needs of rural and semi-urban communities specially for economically poor; production forestry, to cater to the needs of the forest based industries and environmental forestry, which would encompass both in restoring the ecological balance as at present impired.

(e) The National Wasteland Development Board had been set up for dealing with matters relating to development of wasteland in the country, with special emphasis on social forestry-oriented to increase the fuelwood and fodder resources by massive plantation. The function of this Council is policy planning and coordination of all matters relating to the health and scientific management of the land resources of the country.

(f) The life of tribal population is intimately connected with forest and forest economy. For example, in Santhal Pargana forests has been cut recklessly due to stone-breaking work. The trees like Mahua, Sekhua, Herbs, Asan are reduced to minimum. Formerly, they were doing 'Pattal', Datwan making work. Now, they are dependent on contractors for stone-breaking works which has also accelerated the soil erosion, and forced them to work on their own wish. The Government must check these illegal activities and special attention is to be paid to the activities that provide higher incomes to tribals from forest-based activities like production of lac, tassar, silk and other minor forest produce and improve their quality of life.

(g) The National Rural employment programme can be associated with employment-oriented forestry and environmental development programmes as most of the States are doing. In this process degraded and marginal forest areas should receive special attention together with integrated soil and water management.

(h) At the interval of every commissionaries a Regional Forest Research Centre is to be set up. In order to involve people in afforestation by educating them on the importance of trees in the economy and preservation of ecosystem, an extension organisation can also be creased.

(i) The uncontrolled grazing is also playing havoc on naturally regenerated forest crop. In order to contain this problem exclusive pasture can be established everywhere.

(j) There should be an aggressive production programme on all wastelands, marginal and submarginal agricultural lands to meet the needs of fuelwood and small timber. The marginal farmers whose lands are not economically viable can be motivated for tree plantation on the place of traditional cropping.

(k) Smokeless stoves biogas, solar power, wind mills, etc., can be encouraged to ease the pressure on forest.

(l) Universities, colleges can play a vital role in tree planting. In community forestry programme, active participation of school children is must.

(m) Officials should be careful in exotic varieties of tree planting as it can displace indigenous varieties.

(n) The scientific establishment and research agencies should direct their energies towards result-oriented research on improving the ecology like tissue culture for quick multiplication of seedings, work on location-specific species like Jojoba on coastal saline wastelands, redevelopment of the lost mangrove system and aerial seedling in inaccessible areas.

(o) There should be complete moratorium on commercial felling in the Himalayas and other ecologically hilly areas.

(p) Successful use of multi crop pattern in agriculture, the Madhya Pradesh Forest Development Corporation has introduced plantation of more than one species of timber trees, such as bamboo and sheesham in the teak forest in Bari (Hoshangabad). While the teak tree takes about 60 years to mature, the bamboo and sheesham get ready much earlier. Besides, the felling of teak trees too is now planned to be done at three stages—one at the age of 10 and the other at the age of 30—to get better returns.

(q) There should be complete banning of the burning of waste wood in the forest because of creates fire hazards in the summer.

(r) Selection of trees is one of the most important phenomena. If we try to go against nature by planting trees not meant for the areas, nature will revolt and retaliate. In all the degraded area, after a cursery survey of the field, interviews with the local populance about the depleted species, one can decide which species will be better for the area.

(s) To encourage plantation, the people subsisting on selling fuelwood by destroying forest, can be employed to plant trees. The wage should be paid in kind, preferably wheat. It can be stored, if possible, in villages itself to inspire confidence among the people that food is available easily.

(t) To support the massive social forestry programme, adequate research input are needed in respect of fuelwood, biomass, fodder, trees, agro-forestry and genetic engineering. In the field of agro-forestry indepth study is to be done for intercropping relationship between fuelwood and agricultural crops as well as soil/moisture/nutrient/mycorrhizal relationship.

(u) The peoples' participation is must in forestry programme. The meaningful involvement of the people would lead to social fencing resulting the reduction of the unit cost of afforestation.

(v) A huge amount of bio-mass is bunt all over the country for the religious and social purposes such as for dead bodies burning and different festivals like Holi and other local rituals, etc. The electric crematoria can be constructed on mass scale in all over the country and for this people's mentality can be changed through persuasion.

FOREST CLEARANCE IN THE STONE AGE

Perhaps the greatest single step forward in the history of mankind was the transition from hunting to agriculture. In the Mesolithic Age men lived by the spear, the bow and the fishing net; in the Neolithic Age they became farmers. The change came independently at different times in diverse parts of the world. Just how and when men turned to farming in Western Europe has been a subject of debate among naturalists and archaeologists for a hundred years. New methods of dating the implements of Stone Age men have recently given more factual substance to the debate. What is more, we have learned enough about the world in which they lived to test our theories about how they lived by experiment. This is a report of a set of experiments by which a group of scientists in Denmark attempted to re-enact some aspects of the hunting-to-agriculture chapter of mankind's past.

Denmark has unearthed relics of both stages—the bone and flint implements of the Mesolithic hunters and the polished stone axes of the Neolithic farmers. And in ancient lake sediments and bogs the prehistoric tools lie in recognizable strata of pollen, that marvelous dating instrument which identifies each period by its prevailing vegetation. The pollen record, as ecologists read it, tells the following story.

Toward the end of the last ice age, when vegetation was

emerging and the country was still open, hunters ranged all over Denmark. Then. as forests grew dense and reduced large game, men abandoned the forested interior and retreated to the coast, where they made their living by fishing and seal-hunting. This state of affairs continued for thousands of years, until-man suddenly appears in the forest, hacking ou. a new living. Clearings are hewed in the primeval forest. Tree pollen rapidly declines in certain regions, and we find in its place a sharp rise in pollen of herbaceous plants and the emergence of cereals and new weeds, notably plantain—the plant which the Indians of North America called "the footsteps of the white man."

Very shortly a new growth of tree species which typically follow forest clearance—willow, aspen, birch—springs up. The clearence of birch strongly suggests that man used fire to help clear the forest, for on fertile soil birch succeeds a mixed oak forest only after burning. Meanwhile the ground flora undergoes a radical and significant change. Grasses, white clover, sheep sorrel, sheep's-bit and other pasture plants take the upper hand. We can visualize cattle grazing and browsing in grassy meadows bordered by scrub forests of birch and haze!

Finally comes a third phase. The grasses, birch and eventually hazel decline, and a big-tree forest takes over once more. Oak now is more dominant than before; elm and linden never recover the strength they had in the primeval forest.

All this seems to mean that men cleared large areas of the original forest with axes, burned over the clearings, planted small fields of cereals and used the rest for pasturing animals. Their colonization was of short duration; when the forest back, they moved on to clear pollen record, some of their settlements can scarcely have lasted more than 50 years.

Now this is a neat, tidy theory. but there are troublesome questions. Could Neolithic man really have cleared large areas of the thick primieval forest with his crude flint axes? Could he have burned off the felled trees and shrubs in his clearings? Our team of ecologists and archaeologists decided to put these questions to the test of field experiment. We obtained the needed fungus and permission to clear a two-are area in the Draved Forest of Denmark, which is a mixed oak forest like that of Neolithic times.

Two archaeologists, Jorgen Troels-Smith and Svend Jorgensen, took charge of the axe tests. They were able to obtain a number of Neolithic flint axe blades from the National Museum in Copenhagen, and a model for the wooden haft was available in the form of the famous Sigerslev hafted axe excavated from a Danish bog. In Neolithic axes, whose hafts of ash wood, the blade was inserted in a rectangular hole in the haft. Jorgensen and Troels-Smith demonstrated that if the haft was not to be split, it must not hold the blade too tightly but must leave room for a little sidewise play of the blade when it struck.

After making a number of hafted axes fitted with Stone Age man's blades, the two archaeologists, together with two professional lumberjacks, went forth into the forest in September, 1952. When the party attacked the trees, it soon became apparent that the usual tree-chopping technique, in which one puts his shoulders and weight into long, powerful blows, would not do. It often shattered the edge of the delicate flint blade or broke the blade in two. The lumberjacks, unable to change their habits, damaged several axes. The archaeologists soon discovered that the proper way to use the flint axe was to chip at the tree with short, quick strokes, using mainly the elbow and wrist. Troels-Smith, working with an axe blade which had not been sharpened since the Stone Age, employed it effectively throughout the whole clearing operation its without damaging it.

When the two archaeologists reached peak form, they, were able to fell oak trees more than a foot in diameter within half an hour. Small trees they dropped by cutting all around the trunk; on substantial-sized ones they used the slower method of hewing through notches on the opposite sides, in order to control the direction of fall. We realized that for clearing purposes it would be advantageous to have all the trunks lying in the north-south direction: for example, the wood would dry more quickly.

In this manner we cleared the two acres of forest, letting the largest trees stand but killing them by cutting rings through the sapwood. Troels-Smith and Jorgensen concluded that Neolithic men could have cut large clearing in the forests with their flnit axes without great difficulty.

The next problem was to learn how they might have burned off their clearings. For help in this phase of the experiment we called on Kustaa Vilkuna of the University of Helsinki, who is an expect on primitive burning techniques which were still being used quite recently by farmers in the spruce forests of Finland.

Without waiting for the wood to dry, we first tested two burning methods, one modern, the other primitive. The modern method, though effective in forests of conifers, failed completely in our deciduous forest. The primitive method, however, was successful, and we produceeded to use it in the clearing in May of 1954, after the felled trees had, had more than a year to dry. Brushwood and branches cut from the trees were spread over the area to be burned. Then this material was ignited along a 30-foot-wide belt by means of torches of burning birch bark attached to stakes. When the belt was well cleared, we pushed its still burning logs forward with long poles to set fire to the adjacent area. In this way we burned off the tangle of felled vegetation belt by belt. The fire was controlled carefully, day and night, to achieve an even and thorough burning of the ground. It was rather hard work, as oak wood burns slowly, but there were no serious difficulties, and in three or four days the job was finished. We burned only half of the two-acre clearing, because we wished to compare the subsequent growth on burned and unburned ground.

Immediately after the burning we sowed part of the area

with primitive varieties of wheat (einkorn and emmer) and naked barley. That these cereals were grown in Denmark by Neolithic man is shown by grain impressions on excavated pottery. Axel Steensberg, an expert on agricultural methods, old and new, obtained seeds of the cereals from botanical collections and directed our agriculture.

We spread the seeds on the ground, raked them in with a forked branch, and waited for the harvest. For comparison we sowed two sets of plots—one burned and one unburned but hoed and weeded. The contrast in results was remakable. On the unburned ground the grain scarcely grew at all. Evidently the rather acid forest soil was not suited to cerea growing. But the burned ground produced a luxuriant crop (which Steensberg harvested, in Neolithic fashion with a flint knife and a flint sickle). The success of the cereals in this ground was due in part to sweetening of the soil by the wood-ash and the absence of competition from other vegetation, but the burning may also have created other beneficial factors, and we are now in vestigating this matter. In any case whatever the factors are, they are shortlived, for the second year the burned plots yielded much smaller crops.

Now, two years after the clearing and burning, we are in the process of watching developments in the early recovery of natural plant growth. The burned and unburned areas are developing quite differently.

In the area cleared of trees but unburned, events are following an unsurprising and unexciting course. The ground vegetation consists mainly of the species that grew there before the clearing, though it is growing more luxuriously because it has more sunlight. Bracken (ferns), always abundant in this part of the world, is flourishing far more richly than when it was shaded. Grasses and sedges have increased.

The burned ground, on the other hand, is a scene of botanical revolution. Bracken is coming back here too, but most of the other old plants, having shallower roots, were killed off by fire. In their stead we have a whole garden of new plants. Plantain has made its appearance, just as it does in the ancient pollen record after forest clearance. There is a profusion of members of the family Compositae, including dandelions, daisies, sow thistle and so forth. (These plants do not bulk large in the fossil pollen record, but that is understandable because they are pollinated by insects rather than by the wind).

A particularly interesting development is the sudden appearance of mosses and their spread over large patches of the burned area. The main species have never been seen in this forest before. Their spores have flown into the clearing on the wind, and no doubt mosses came the same way to the areas burned by Neolithic man. What makes them especially significant is that certain mosses seem to be definite indicators of fire; three species have been so identified in America, and sure enough the same three appeared in our burned clearing. Since the moss phase in a burned forest must be ephemeral, moss spores in the fossil record should enable us to pinpoint the dates of clearance by Neolithic man and to learn whether they burned the same clearing more than once during the existence of a continuous settlement. Unfortunately the small moss spores are difficult to recognize, and analysts of the ancient pollen deposits have not counted them hitherto. We made a small test count at the site of a Neolithic forest clearing in Denmark, analyzing the layers representing the time of the clearance and the period just before. According to our fragmentary count, there was a sharp rise in general moss growth (we made no attempt to distinguish individual species) immediately after the clearance of the area.

Our experimental clearing in the Danish forest is just beginning to pass into the second phase, when pioneer trees appear and the regeneration of the forest commences. Birch seedlings are starting to spring up in profusion; willow seedlings have appeared; and hazel, aspen and linden shoots are rising from roots that were not killed by the fire. We are looking forward to studying this gradual regeneration in the years to come, as well as to reliving the stage in Neolithic farming when men grazed their cattle on the re-emerging ground vegetation.

Meanwhile we can say that so far our experiment has confirmed the archaeological interpretation of the pollen record on several important counts. It has been demonstrated that the forest could indeed have been cleared by the primitive tools of Neolithic man, and that in the first stage at least the reviving vegetation follows a course very like that deduced from the ancient pollen layers.

Of course man's transition from hunting to farming may well have taken other paths besides the one we have traced in the Danish clearings. More than one type of agriculture may have existed simultaneously in Denmark. As a matter of fact, Troels-Smith has found evidences of a more primitive agriculture during the same period on the Danish coast, where the Middle Stone Age men apparently cleared no forests but practiced a little crude farming along with their hunting and fishing.

The Neolithic farming culture described in this article is so much more advanced, and begins so suddenly, that it seems to signal the arrival and invasion of a vigorous new people from another region.

14

Toxic Substances and Ecological Cycles

The vastness of the earth has fostered a tradition of unconcern about the release of toxic wastes into the environment. Billowing clouds of smoke are diluted to apparent nothingness; discarded chemicals are flushed away in rivers; insecticides "disappear" after they have done their job; even the massive quantities of radioactive debris of nuclear explosions are diluted in the apparently infinite volume of the environment. Such pollutants are indeed diluted to traces—to levels infinitesimal by ordinary standards, measured as parts per billion or less in air, soil and water. Some pollutants do disappear; they are immobilized or decay to harmless substances. Others last, sometimes in toxic form, for long periods. We have learned in recent years that dilution of persistent pollutants even to trace levels detectable only by refined techniques is no guarantee of safety. Nature has ways of concentrating substances they are frequently surprising and occasionally disastrous.

We have had dramatic examples of one of the hazards in the dense smogs that blanket our cities with increasing frequency. What is less widely realized is that there are global, long-term ecological processes that concentrate toxic substances, sometimes hundreds of thousands of times above levels in the environment. These processes include not only patterns of air and water circulation but also a complex series of biological mechanisms. Over the past decade detailed studies of the distribution of both radioactive debris and pesticides have revealed patterns that have surprised even biologists long familiar with the unpredictability of nature.

Major contributions to knowledge of these patterns have come from studies of radioactive fallout. The incident that triggered worldwide interest in large-scale radioactive pollution was the hydrogen-bomb test at Bikini in 1954 known as "Project Bravo." This was the test that inadvertently dropped radioactive fallout on several Pacific islands and on the Japanese fishing vessel Lucky Dragon. Several thousand square miles of the Pacific were contaminated with fallout radiation that would have been lethal to man. Japanese and U.S. oceanographic vessels surveying the region found that the radioactive debris had been spread by wind and water, and, more disturbing, it was being passed rapidly along food chains from small plants to small marine organisms that ate them to larger animals (including the tuna, a staple of the Japanese diet).

The U.S. Atomic Energy Commission and agencies of other nations, particularly Britain and the U.S.S.R., mounted a large international research programme, costing many millions of dollars, to learn the details of the movement of such debris over the earth and to explore its hazards. Although these studies have been focused primarily on radioactive materials, they have produced a great deal of basic information about pollutants in general. The radioactive substances serve as tracers to show the transport and concentration of materials by wind and water and the biological mechanisms that are characteristic of natural communities.

One series of investigations traced the worldwide movement of particles in the air. The tracer in this case was strontium 90, a fission product released into the earth's atmosphere in large quantities by nuclear-bomb tests. Two reports in 1962—one by S. Laurence Kulp and Arthur R. Schulert of Columbia University and the other by a United Nations committee—furnished a detailed picture of the travels of strontium 90. The isotope was concentrated on the ground between the latitudes of 30 and 60 degrees in both hemispheres, but concentrations were five to 10 times greater in the Northern Hemisphere, where most of the bomb tests were conducted.

It is apparently in the middle latitudes:

Forest Community is an integrated array of plants and animals that accumulates and reuses nutrients in stable cycles, as indicated schematically in black. DDT participates in parallel cycles (colour). The author measured DDT residues in a New Brunswick forest in which four pounds per acre of DDT had been applied over seven years. (Studies have shown about half of this landed in the forest. the remainder dispersing in the atmosphere.) Three years after the spraying, residues of DDT were as shown (in pounds per acre).

that exchanges occur between the air of upper elevations (the stratosphere) and that of lower elevations (the troposphere). The larger tests have injected debris into the stratosphere; there it remains for relatively long periods, being carried back into the troposphere and to the ground in the middle latitudes in late winter or spring. The mean "half-time" of the particles' residence in the stratosphere (that is, the time for half of a given injection to fall out) is from three months to five years, depending on many factors, including the height of the injection, the size of the particles, the latitude of injection and the time of year. Debris injected into the troposphere has a mean half-time of residence ranging from a few days to about a month. Once airborne, the particles may travel rapidly and far. The time for one circuit around the earth in the middle latitudes varies from 25 days to less than 15, (Following two recent bomb tests in China fallout was detected at the Brookhaven

National Laboratory on Long Island respectively nine and 14 days after the tests.)

Numerous studies have shown further that precipitation (rain and snowfall) plays an important role in determining where fallout will be deposited. Lyle T. Alexander of the Soil Conservation Service and Edward P. Hardy, Jr., of the AEC found in an extensive study in Clallam County, Washington, that the amount of fallout was directly proportional to the total annual rainfall.

It is reasonable to assume that the findings about the movement and fallout of radioactive debris also apply to other particles of similar size in the air. Thus conclusion is supported by a recent report by Donald F. Gatz and A. Nelson Dingle of the University of Michigan, who showed that the concentration of pollen in precipitation follows the same pattern as that of radioactive fallout. This observation is particularly meaningful because pollen is not injected into the troposphere by a nuclear explosion; it is picked up in air currents from plants close to the ground. There is little question that dust and other particles, including small crystals of pesticides, also follow these patterns.

From these and other studies it is clear that various substances released into the air are carried widely around the world and may be deposited in concentrated form far from the original source. Similarly. most bodies of water—especially the oceans—have surface currents that may move materials five to 10 miles a day. Much higher rates, of course, are found in such major oceanic currents as the Gulf Stream. These currents are one more physical mechanism that can distribute pollutants widely over the earth.

The research programmes of the AEC and other organizations have explored not only the pathways of air and water transport but also the pathways along which pollutants are distributed in plant and animal communities. In this connection we must examine what we mean by a "community."

Biologists define communities broadly to include all species, not just man. A natural community is an aggregation of a great many different kinds of organisms, all mutually interdependent. The basic conditions for the integration of a community are determined by physical characteristics of the environment such as climate and soil. Thus a sand dune supports one kind of community. a freshwater lake another, a high mountain still another. Within each type of environment there develops a complex of organisms that in the course of evolution becomes a balanced, self-sustaining biological system.

Such a system has a structure of interrelations that endows the entire community with a predictable developmental pattern, called "succession," that leads toward stability and enables the community to make the best use of its physical environment. This entails the development of cycles through which the community as a whole shares certain resources, such as mineral nutrients and energy. For example, there are a number of different inputs of nutrient elements into such a system. The principal input is from the decay of primary minerals in the soil. There are also certain losses, mainly through the leaching of substances into the underlying water table. Ecologists view the cycles in the system as mechanisms that have evolved to conserve the elements essential for the survival of the organisms making up the community.

One of the most important of these cycles is the movement of nutrients and energy from one organism to another along the pathways that are sometimes called food chains. Such chains start with plants, which use the sun's energy to synthesize organic matter; animals eat the plants; other animals eat these herbivoures, and carnivores in turn may constitute additional levels feeding on the herbivoures and on one another. If the lower orders in the chain are to survive and endure. there must be a feed-back of nutrients. This is provided by decay organisms (mainly microorganisms) that break down organic debris into the substances used by plants. It is also obvious that the community will not survive if essential links in the chain are eliminated; therefore the preying of one level on another must be limited.

Ecologists estimate that such a food chain allows the transmission of roughly 10 per cent of the energy entering one level to the next level above it, that is, each level can pass on 10 per cent of the energy it receives from below without suffering a loss of population that would imperil its survival. The simplest version of a system of this kind takes the form of a pyramid, each successively higher population receiving about a tenth of the energy received at the level below it.

Actually nature seldom builds communities with so simple a structure. Almost invariably the energy is not passed along in a neatly ordered chain but is spread about to a great variety of organisms through a sprawling, complex web of pathways. The more mature the community, the more diverse its makeup and the more complicated its web. In a natural ecosystem the network may consist of thousands of pathways.

This complexity is one of the principal factors we must consider in investigating how toxic substances may be distributed and concentrated in living communities. Other important basic factors lie in the nature of the metabolic process. For example, of the energy a population of organisms receives as food, usually less than 50 per cent goes into the construction of new tissue, the rest being spent for respiration. This circumstance acts as a concentrating mechanism: a substance not involved in respiration and not excreted efficiently may be concentrated in the tissues twofold or more when passed from one population to another.

Let us consider three types of pathway for toxic substances that involve man as the ultimate consumer. The three examples, based on studies of radioactive substances, illustrate the complexity and variety of pollution problems.

The first and simplest case is that of strontium 90. Similar to calcium in chemical behaviour. this element is concentrated in bone. It is a long-lived radioactive isotope and is a hazard because its energetic beta radiation can damage the mechanisms involved in the manufacture of blood cells in the bone marrow. In the long run the irradiation may produce certain types of cancer. The route of strontium 90 from air to man is rather direct: we ingest it in leafy vegetables, which absorbed it from the soil or received it as fallout from the air, or in milk and other dairy products from cows that have fed on contaminated vegetation. Fortunately strontium is not usually concentrated in man's food by an extensive food chain. Since it lodges chiefly in bone, it is not concentrated in passing from animal to animal in the same ways other radioactive substances may be (unless the predator eats bones!).

Quite different is the case of the radioactive isotope cesium 137. This isotope, also a fission product, has a long-lived radioactivity (its half-life is about 30 years) and emits penetrating gamma rays. Because it behaves chemically like potassium, an essential constituent of all cells, it becomes widely distributed once it enters the body. Consequently it is passed along to meat-eating animals, and under certain circumstances it can accumulate in a chain of carnivores.

A study in Alaska by Wayne C. Hanson, H.E. Palmer and B.I. Griffin of the AEC's Pacific-Northwest Laboratory showed that the concentration factor for cesium 137 may be two or three for one step in a food chain. The first link of the chain in this case was lichens growing in the Alaskan forest and tundra. The lichens collected cesium 137 from fallout in rain. Certain caribou in Alaska live mainly on lichens during the winter, and caribou meat in turn is the principal diet of Eskimos in the same areas. The investigators found that caribou had accumulated about 15 micromicrocuries of cesium radioactivity per gram of tissue in their bodies. The Eskimos who fed on these caribou had a concentration twice as high (about 30 micromicrocuries per gram of tissue) after eating many pounds of caribou meat in the course of a season. Wolves and foxes that ate caribou sometimes contained three times the concentration in the flesh of the caribou. It is easy to see that in a longer chain, involving not just two animals but several, the concentration of a substance that was not excreted or metabolized could be increased to high levels.

A third case is that of iodine 131, another gamma ray emitter. Again the chain to man is short and simple: The contaminant (from fallout) comes to man mainly through cows' milk, and thus the chain involves only grass, cattle, milk and man. The danger of iodine 131 lies in the fact that iodine is concentrated in the thyroid gland. Although iodine 131 is short-lived (its half-life is only about eight days), its quick and localized concentration in the thyroid can cause damage. For instance, a research team from the Brookhaven National Laboratory headed by Robert Conard has discovered that children on Rongelap Atoll who were expased to fallout from the 1954 bomb test later developed thyroid nodules.

The investigations of the iodine 131 hazard yielded two lessons that have an important bearing on the problem of pesticides and other toxic substances released in the environment. In the first place we have had a demonstration that the hazard of the toxic substance itself often tends to be underestimated. This was shown to be true of the exposure of the thyroid to radiation. Thyroid tumors were found in children who had been treated years before for enlarged thymus glands with doses of X rays that had been considered safe. As a result of this discovery and studies of the effects of iodine 131, the Federal Radiation Council in 1961 issued a new guide reducing the permissible limit of exposure to ionizing radiation to less than a tenth of what had previously been accepted. Not the least significant aspect of this lesson is the fact that the toxic effects of such a hazard may not appear until long after the exposure; on Rongelap Atoll 10 years passed before the thyroid abnormalities showed up in the children who had been exposed .

The second lesson is that, even when the pathways are well understood, it is almost impossible to predict just where toxic substances released into the environment will 'reach dangerous levels. Even in the case of the simple pathway followed by iodine 131 the eventual destination of the substance and its effects on people are complicated by a great many variables: the area of the cow's pasture (the smaller the area, the less fallout the cow will pick up); the amount and timing of rains on the pasture (which on the one hand may bring down fallout but on the other may wash it off the forage); the extent to which the cow is given stored, uncontaminated feed the amount of iodine the cow secretes in its milk; the amount of milk in the diet of the individual consumer, and so on.

If it is difficult to estimate the nature and extent of the hazards from radioactive fallout, which have been investigated in great detail for more than a decade by an international research programme, it must be said that we are in a poor position indeed to estimate the hazards from pesticides. So far the amount of research effort given to the ecological effects of these poisons has been comparatively small, although it is increasing rapidly. Much has been learned, however, about the movement and distribution of pesticides in the environment, thanks in part to the clues supplied by the studies of radioactive fallout.

Our chief tool in the pesticide inquiry is DDT. There are many reasons for focusing on DDT; it is long-lasting, it is now comparatively easy to detect, it is by far the most widely used pesticide and it is toxic to a broad spectrum of animals, including man. Introduced only a quarter-century ago and spectacularly successful during World War II in controlling body lice and therefore typhus, DDT quickly became a universal weapon in agriculture and in public health campaigns against disease-carriers. Not surprisingly, by this time DDT has thoroughly permeated our evironment. It is found in the air of cities, in wildlife all over North America and in remote corners of the earth, even in Adelie penguins and skua gulls (both carnivores) in the Antarctic. It is also found the world over in the fatty tissue of man. It is fair to say that there are probably few populations in the world that are not contaminated to some extent with DDT.

We now have a considerable amount of evidence that DDT is spread over the earth by wind and water in much the same patterns as radioactive fallout. This seems to be true in spite of the fact that DDT is not injected high into the atmosphere by an explosion. When DDT is sprayed in the air, some fraction of it is picked up by air currents as pollen is, circulated through the lower troposphere and deposited on the ground by rainfall. I found in tests in Marine and New Brunswick, where DDT has been sprayed from airplanes to control the spruce budworm in forest, that even in the open, away from trees, about 50 per cent of the DDT does not fall to the ground. Instead it is probably dispersed as small crystals in the air. This is true even on days when the air is still and when the low-flying planes release the spray only 50 to 100 feet above treetop level. Other mechanisms besides air movement can carry DDT for great distances around the world. Migrating fish and birds can transport it thousands of miles. So also do oceanic currents. DDT has only a low solubility in water (the upper limit is about one part per billion), but as algae and other organisms in the water absorb the substance in fats, where it is highly soluble, they make room for more DDT to be dissolved into the water. Accordingly water that never contains more than a trace of DDT can continuously transfer it from deposits on the bottom to organisms.

DDT is an extremely stable compound that breaks down very slowly in the environment. Hence with repeated spraying the residues in the soil or water basins accumulate. Working with Frederic T. Martin of the University of Marine, I found that in a New Brunswick forest where spraying had been discontinued in 1958 the DDT content of the soil increased from half a pound per acre to 1.8 pounds per acre in the three years between 1958 and 1961. Apparently the DDT residues were carried to the ground very slowly on foliage and decayed very little. The conclusion is that DDT has a long half-life in the trees and soil of a forest, certainly in the range of tens of years.

Doubtless there are many places in the world where reservoirs of DDT are accumulating. With my colleagues Charles F. Wurster, Jr., and Peter A, Isaacson of the State University of New York at Stony Brook, I recently sampled a marsh along the south shore of Long Island that had been sprayed with DDT for 20 years to control mosquitoes. We found that the DDT residues in the upper layer of mud in this marsh ranged up to 32 pounds per acre!

We learned further that plant and animal life in the area constituted a chain that concentrated the DDT in spectacular fashion. At the lowest level the plankton in the water contained .04 part per million of DDT; minnows contained one part per million, and a carnivorous scavenging bird (a ring-billed gull) contained about 75 parts per million in its tissues (on a whole-body, wet-weight basis). Some of the carnivorous animals in this community had concentrated DDT by a factor of more than 1,000 over the organisms at the base of the ladder.

A further tenfold increase in the concentrations along this food web would in all likelihood result in the death of many of the organisms in it. It would then be impossible to discover why they had disappeared. The damage from DDT concentration is particularly serious in the higher carnivores. The mere fact that conspicuous mortality is not observed is no assurance of safety. Comparatively low concentrations may inhibit reproduction and thus cause the species to fade away.

That DDT is a serious ecological hazard was recognized from the beginning of its use. In 1946 Clarence Cottam and Elmer Higgins of the U.S. Fish and Wildlife Service warned in the Journal of Economic Entomology that the pesticide was a potential menance to mammals, birds, fishes and other wildlife and that special care should be taken to avoid its application to streams, lakes and coastal bays because of the sensitivity of

Location		*Organism*	*Tissue*	*Concentration (Parts Per Million)*
U.S. (Average)		Man	Fat	11
Alaska (Eskimo)				2.8
England				2.2
West Germany				2.3
France				5.2
Canada				5.3
Hungary				12.4
Israel				19.2
India				12.8-31.0
U.S.	California	Plankton		5.3
	California	Bass	Edible Flesh	4-138
	California	Grebes	Visceral Fat	Up to 1,600
	Montana	Robin	Whole Body	6.8-13.9
	Wisconsin	Crustacea		.41
	Wisconsin	Chub	Whole Body	4.52
	Wisconsin	Gull	Brain	20.8
	Missouri	Bald Eagle	Eggs	1.1-5.6
	Connecticut	Osprey	Eggs	6.5
	Florida	Dolphin	Blubber	About 220
Canada		Woodcock	Whole Body	1.7
Antarctica		Penguin	Fat	.015-18
Antarctica		Seal	Fat	.042-.12
Scotland		Eagle	Eggs	1.18
New Zealand		Trout	Whole Body	6-8

DDT Residues, which include the derivatives DDD and DDE as well as DDT itself, have adparently entered most food webs. These data were selected from hundred of reports that show DDT has a worldwide distribution, with the highest concentrations in carnivorous birds.

fishes and crabs. Because of the wide distribution of DDT the effects of the substance on a species of animal can be more damaging than hunting or the elimination of a habitat (through an operation such as dredging marshes). DDT affects the entire species rather than a single population and may well wipe out the species by eliminating reproduction.

Within the past five years, with the development of improved techniques for detecting the presence of pesticide residues in animals and the environment, ecologists have been able to measure the extent of the hazards presented by DDT and other persistent general poisons. The picture that is emerging is not a comforting one. Pesticide residues have now accumulated to levels that are catastrophic for certain animal populations, particularly carnivorous birds. Furthermore, it has been clear for many years that because of their shotgun effect these weapons not only attack the pests but also destroy predators and competitors that normally tend to limit proliferation of the pests. Under exposure to pesticides the pests tend to develop new strains that are resistant to the chemicals. The result is an escalating chemical warfare that is self-defeating and has secondary effects whose costs are only beginning to be measured. One of the costs is wildlife, notably carnivorous and scavenging birds such as hawks and eagles. There are others; destruction of food webs aggravates pollution problems, particularly in bodies of water that receive mineral nutrients in sewage or in water draining from heavily fertilized agricultural lands. The plant populations, no longer consumed by animals, fall to the bottom to decay anaerobically, producing hydrogen sulphide and other noxious gases, further degrading the environment.

The accumulation of persistent toxic substances in the ecological cycles of the earth is a problem to which mankind will have to pay increasing attention. It affects many elements of society, not only in the necessity for concern about the disposal of wastes but also in the need for a revolution in pest control. We must learn to use pesticides that have a short half-life in the environment—better yet, to use pest-control techniques that do not require applications of general poisons. What has been learned about the dangers in polluting ecological cycles is ample proof that there is no longer safety in the vastness of the earth.

15

Predatory Fungi

Unless man succeeds in duplicating the process of photosynthesis, it appears that animals will always have to feed upon plants. But the plant world exacts its retribution. A number of plants have turned the tables on the animal kingdom, reversing the roles of predator and prey. These are the plant carnivores—plants that trap and consume living amials. Most famous are the pitcher plant, with its reservoir of digestive fluid in which to drown hapless insects; the sundew, with its fly-paper-like leaves; and Venus's-flytrap, with its snapping jaws. But there are other carnivorous plants of large significance in the balance of nature. We ought to know them better because they are to be found in great profusion and variety in any pinch of forest soil or garden compost. They are microscopic in size, but just as deadly their animal prey as the sundew or Venus's flytrap.

These tiny predators are members of the large group of fungi we call molds. They grow in richly branching networks of filaments visible to the naked eye as hairy or velvety mats. Molds do not engage in photosynthesis. Like most bacteria, they lack chlorophyll and so must derive their food from other plants and from animals. Molds have long been familiar as scavengers of dead organisms, promoters of the process of decay. It was not until 1888 that a German mycologist, named Friedrich Wilhelm Zopf, beheld molds in the act of trapping and killing live animals—in this case the larvae of a tiny worm, the wheat-cockle nematode.

The nematodes (eelworms, hook-worms and their like) are not the only prey of these animal-eating plants. Their victims run the gamut from the comparatively formidable nematodes down to small crustaceans, rotifers and the lowly amoeba. Charles Drechsler of the U.S. Department of Agriculture, a student of the subject for some 25 years, has identified a large number of carnivorous molds and matched them to their prey. Many are adapted to killing only one species of animal, and some are equipped with traps and shares which are marvels of genetic resourcefulness, How they evolved their predatory habits and oragans remains an evolutionary mystery, These molds belong to quite different species and have in common only their behaviour and some similarities of trapping technique. They present a challenging subject for investigation which may throw light on some fundamental questions in biology and may lead also to new methods for control of a number of crop-killing nematodes.

The simplest of the molds have no special organs with which to ensnare their victims. Their filaments, however, secrete a sticky substance which holds fast any small creature that has the misfortune to come in contact with it. The mold then injects daughter filaments into the body cavity of the victim and digests its contents. Most of the animals caught in this way are rhizopods—sluggish amoebae encased in minute hard shells. Sometimes, however, the big, vigorous soil nematodes are trapped by this elementary means.

More specialized is an unusual water mold, of the genus Sommerstcraffia which catches rotifers, its actively swimming prey, with little sticky pegs that branch from its filaments. When a rotifer, browsing among the algae on which this mold grows, takes one of these pegs in its ciliated mouth, it finds itself impaled like a fish on a hook.

Some molds do their trapping in the spore stage. The parent mold produces staggering numbers of sticky spores. When a spore is swallowed by or sticks to a passing amoeba or nematode, it germinates in the body of its luckless host and sends forth from the shriveled corpse new filaments and new spores to intercept other victims.

The most remarkable of all killer molds are found among the so-called Fungi Imperfecti, or completely asexual fungi, The advanced specialization of these molds is particularly interesting because they are not killers by obligation but can live quite well on decaying organic matter when nematodes, their animal victims, are not available. If nematodes are present, these molds immediately develop highly specialized structures which re-adapt them to a carnivorous way of life. They will do so even if they are merely wetted with water in which nematodes have lived.

One of these molds is Arthrobotrys oligospora, the nematode-catching fungus that was first studied by Zopf. When nematodes are available, it develops net-works of loops, fused together to form an elaborate nematode trap. An exteremely sticky fluid secreted by the mold seems to play an important role in capturing the nematode, which need not even enter the network in order to be held fast. The fluid is so sticky that one-point contact with the network frequently is enough to doom the nematode. In its frenzied struggles to escape, the worm only becomes further entangled in the loops, and finally, after a few hours of exertion, weakens and dies. The destruction these molds can cause in a laboratory culture of nematodes is appalling, particularly from the point of view of the nematode!

Two French biologist, Jean Comandon and Pierre de Fonbrune, have made motion pictures which show that the fungus's secretion of this adhesive substance is accompanied by intense activity in its cells. Material in the cytoplasm of the cell streams toward the point of contact with the worm. The mold may be bringing up reserves of adhesive and digestive enzymes to subdue the nematode; it may also secrete a narcotic or an intoxicant to speed the process.

Even more artfully contrived are the "rabbit snares" employed by some molds. First fully described by Charles Drechsler, these are rings of filament which are attached by short branches to the main filaments, hundreds of them growing on one mold plant. The rings are always formed by three cells and have an inside diameter just about equal to the thickness of a nematode. When a nematode in its blind wanderings through the soil, has the ill luck to stick its head into one of these rings, the three cells suddenly inflate like a pneumatic tire, gripping it in a stranglehold from which there is no escape.

The rings respond almost instantaneously to the presence of a nematode; in less than one tenth of a second the three cells expand to two or three times their former volume, obliterating the opening of the ring. It is difficult to understand how the delicate filaments can hold the powerfully thrashing worm in so unyielding a grip. Occasionally a muscular worm does escape by breaking the ring off its stalk. But this victory only postpones the inevitable. The ring hungs on like a deadly collar and ultimately generates filaments which invade the worm, kill it and consume it.

We are not yet sure what cellular mechanisms activate these deadly nooses. We know that in the case of the constricting ring of one mold the activating stimulus is the sliding touch of the nematode as it enters the ring. A nematode that touches the outer surface of the ring will not trigger the mechanism. But if the worm passes inside the ring, its doom is certain. This mold, then, exhibits a sharply localized "paratonic" or touch response like that of the Venus's-flytrap.

Perhaps the inflation of the cells is caused by a change in osmotic pressure, resulting in an intake of water either from the environment or from neighbouring cells. Or perhaps it results from changes in the colloidal structure of the cell protoplasm There is a recent report from England that the constricting rings of one species react to acetylcholine—the substance associated with the transmission of impulses across synapses in the animal nervous system!

Some species of molds prey upon root eelworm that infest cereal crops, potatoes and pineapples. This has inspired experiments to use these fungi to control the pests. In one early experiment, conducted in Hawaii by M. B. Linford of the University of Illinois and his associate, a mulch of chopped pine-apple tops was added to soil known to harbour the pine-apple root-knot eelworm. This mulch produced an increase in the numbers of harmless, free-wandering nematodes which thrive in rich soil. The presence of these decay nematodes stimultated molds in the soil to develop nematode traps, which caught the eelworms as well as the harmless species. A recent experiment in England gives similar promise that the molds may be effective against the cereal root eelworm. Plants protected by stimulated molds showed slight damage compared to the eelwormravaged control plants.

Investigators in France have reported an experiment which suggests that molds may be used to control nematode parasites of animals as well as those of plants. To sheep pens were heavily infested with larvae of a hookworm, closely related to the hookworms of man, which causes severe pulmonary and intestinal damage to sheep. One of the pens was sprinkled with the spores of three molds that employ snares or sticky nets to trap nematodes. Healthy lambs were placed in both pents. After 35 days of exposure the lambs in the pen inoculated with the molds were found free of infection, while those in the control pen showed signs of infestation with the worm.

The carnivorous molds offer many possibilities for future investigation. One subject that needs to be explored is their role in the complex biology of the soil. We would also like to know more about the physiological mechanism that underlies the extraordinary behaviour of the nematode "snares." The results of experiments of mold control of nematodes are already encouraging. They suggest that one day these peculiar little plants may perform an even more important role in aggricul-ture than they played in nature, silently and unobtrusively, throughout the millennia before their discovery.

16

Human Impact on World Ecosystems

Determining the full effect of human activities on the environment is a difficult task. One reason for this is its magnitude: the amounts of food, fuel, and other raw materials that humans consume; the physical destruction of landscapes and natural ecosystems that they accomplish; and the volume of wastes, many highly toxic, that they discharge into the environment are so enormous as to tax our capacity to measure them. The task is further complicated by the fact that modern technology is producing environmental changes with unprecedented and unanticipated speed. Moreover, the toxic effects of many harmful wastes do not become evident until the environment has been exposed to them for 10 to 20 years.

Although some human activities benefit the environment, their net effect has been its increasingly rapid deterioration, a serious matter in that, as aerobic heterotrophs, humans are an integral part of the ecosystem in which they live. We will forever depend for our good health and existence upon an environment that is relatively free of lethal and disease-producing poisons and that will provide us with fresh, oxygenrich air; pure drinking water; a large variety of pure foods; and raw biological materials required for clothing, shelter, and other necessities.

This chapter describes some of the kinds of damage that humans have done and are doing to their environment, some of the ways in which the lessening of environmental quality threatens the existence of earth's organisms, and some of the means that are or can be employed to arrest environmental deterioration.

EXHAUSTION OF NATURAL RESOURCES

Utilization of natural resources has enormous impact on environmental quality in several ways. Overfishing, overgrazing, ard overlumbering, for example, have often so severely depleted natural populations that they can no longer serve as sources of fish, forage, or wood. Moreover, depletion of their populations lessens the ability of natural ecosystems to recycle air and water and thus purify them of toxic wastes. The use of mineral resources can also adversely affect environmental quality, not only during mining and processing, but also in the use of the products derived from them. For example, discarded mine residues, or tailings, often produce acid runoffs that poison lakes and streams; offshore oil well blowouts and tanker wrecks pollute the sea; and processing plants produce toxic wastes that contaminate water supplies. As deposits of a particular mineral become depleted, new technologies for extracting the mineral from lower-grade ores have often proved to be more polluting than traditional methods. The use of manufactured products, such as insecticides and fuels, also often pollutes air, soil, and water. Unfortunately, examples of the adverse effects of resource utilization on ecosystems are numberous.

Fossil Fuels

Coal, natural gas and crude oil, or petroleum, the major fossil fuels, are all derived from organic matter. Most coal was formed during the Pennsylvanian period, 320 to 280 million years ago, from buried vegetation of the great club moss and free fern swamps of that time. Petroleum is apparently composed of the decomposition products of marine phyto-and zooplankton trapped in the bottom muds of inland seas. Because of their porosity, sandstone and limestone sedimentary rocks* are the principal reservoirs of petroleum.

Coal. Following World War II, consumption of coal, once the most widely used fossil fuel, dropped to second and then third place among fossil fuels in many industrialized countries as oil and then natural gas became more readily available. As oil and gas reserves decline, many expect coal to again become the predominant fuel of industrialized nations. Mining and use of coal, however, have usually exacted a high cost in air and water pollution—including that by acid rain—and landscape destruction by strip mining (open pit mining) and similar practices, in addition to the diseases and accidents associated with mining.

Crude Oil. While the United States has enough coal for several hundred years of expanded energy use, like most oil-producing countries, it is rapidly depleting its crude oil reserves. Although it is the world's third largest producer of oil, the United States is estimated to have only about a 20-year supply of readily extractable oil. The world's known reserves of crude oil are expected to be depleted by the year 2010. Because the world has come to depend so heavily on crude oil, not only for fuel but for such products as lubricants, fertilizers, and plastics, we have become especially conscious of these dwindling petroleum reserves. New interest has also developed in synthetic fuels, or synfuels, such as gasified coal, coal which has been converted to methane or other fuel.

Oil Sands and Oil Shale. Very large deposits of oil sands exist in Canada and some western areas of the United States. Because crude oil is less expensive to obtain and process than oil extracted from sands, this resources has not yet been exploited. Oil shale is a solid that, theoretically, can be profitably processed to separate its gas, oil, and various other fractions, such as sulphur, Oil shale deposits in the United States are 20 times larger than the country's reserves of oil sands. Despite continuing efforts, however, no satisfactory process has yet been developed to extract shale oil economically. Environmental problems are also involved, the most serious of which is the possibility that the well water and rivers of several western states would be poisoned by by-products of the extraction process. The use of oil and its products as fuel, especially types containing a substantial amount of sulphur, contributes greatly to air and water pollution.

Uranium

Supplies of uranium ore, upon which the operation of conventional nuclear power plants depends, are severely limited. Conventional reactors, as opposed to breeder or fusion reactors,* use uranium-235 (^{235}U) as fuel. However, the most abundant form of uranium in most uranium ores is uranium-238 (^{238}U). Uranium-235 makes up only 0.7 per cent of the ore's uranium fraction. This means that uranium ore must be enriched, or processed to increase its proportion of ^{235}U to a level at which it can sustain a chain reaction and thus be used as reactor fuel.

At the originally intended rate of expansion of the United States nuclear power industry, severe shortages of domestic uranium ore would have occurred by 1990 or 1995 if reliance on conventional reactors continued; virtual exhaustion of ^{235}U would have occurred by 2020. The ^{235}U ore reserves in the United States amount to only 1.5 million tons; with conventional reactors, by the year 2000, some 2.4 million tons would have been used. Without an eventual switch to breeder-reactor technology or an unlikely breakthrough in the perfection of fusion reactors, nuclear power in the United States was to

*A breeder reactor produces more radioactive fuel, in the form of plutonium, than it consumes. A fusion reactor is essentially a harnessed H-bomb in which deuterium nuclei are fused with tritium nuclei.

have had a rather brief history. However, the discovery of several additional world deposits of uranium ore in the early 1980s has extended its life somewhat.

Uranium mining, processing, transport, and use all pose some risks of radioactive pollution of the environment. More serious, however, are the environmental hazards associated with the storage and reprocessing of high level radioactive wastes and the storage of the radioactive components of decommissioned nuclear power plants.

Other Minerals

The question of the adequacy of the world's supply of mineral resources has long been controversial. Many of the minerals used in industry are nonrenewable or can be recycled only at great expense. The total reserves of many valuable minerals are largely unknown. The difficulty of estimating mineral reserves is demonstrated by the fact that known reserves of several elements once thought to be inexhaustible were judged to be dwindling dangerously by the late 1960s and early 1970s. By the late 1970s, however, the discovery of several new sources and the development of new methods of ore extraction had radically improved the outlook for several of these elements. For example, a new process for extracting iron from the ore taconite removed the looming threat of iron shortages in the United States. Also, rich sources of iron and nickel were discovered in the 1960s in Australia and more recently in Brazil. In addition, it was discovered that enormous quantities of manganese, copper, and other minerals are deposited as nodules on the sea floor. Given the present rates of consumption, these nodules contain enough copper to last 6,000 years, enough nicked to last 150,000 years, and enough manganese to last 400,000 years.

For the most part, we have mined the richest ores and are now extracting elements from lower-grade ores through the use of advanced technology. However, such processes usually raise the cost of the elements, use more water and energy than do traditional methods, and increase rates of pollution. Furthermore, the discovery of new sources and the development of new extraction technologies cannot change the fact that the mineral reserves of our planet are finite. One of the best ways to meet increased demands for many minerals, while taking into account their limited supply, is by recycling. This approach reduces pollution by removing discarded items from the environment and by lessening the need for, and the pollution resulting from, mining and processing ore. Although extracting metals from scrap generates its own pollution and requires considerable energy, the energy efficiency of recycling can be dramatic. For example, the production of aluminum from discarded cans saves up to 95 per cent of the energy necessary to extract the same amount of aluminum from ore. Although iron extraction requires large amounts of energy, the production of steel from some scrap iron can save as much as 75 per-cent of the energy cost of extraction from iron ore. Another way to conserve resources, reduce energy use, and abate pollution is to produce goods that are durable or can be repaired.

Topsoil

It is easy to see why we should conserve short supplies of non-renewable resources. Yet slowly renewable resources need to be guarded just as closely. Topsoil, for example, that upper portion of the soil from which the roots of food crops absorb their nutrients, is not only essential for growing such crops but for maintaining natural communities. It accumulates very slowly—in grasslands, at a rate of only a few inches per century. Without proper protection, however, a single storm can wash or blow away a layer of topsoil several inches deep. Most of this soil is then lost to lakes or oceans.

Topsoil Destruction. Although the loss of the earth's topsoil is reaching crisis proportions in many parts of the world, it is by no means a modern problem. From the first gathering of wild crops by primitive nomads, from the first tilling of the soil by ancient farmers, and from the first felling of trees for firewood and building materials, the soil has been abused. Whenever in human history slopes were overgrazed or stands of trees on hillsides were cut or burned away, rains soon began to wash the topsoil into the valleys below.

In moderation, woodcutting and grazing need not destroy a forest or the watershed (drainage area) it covers. However, extensive clear-cutting, burning, and over grazing and any cutting of trees on sleep slopes are invariably destructive unless the slopes are replanted at once. Denuded of trees, a hillside no longer retains rainwater, and lowland springs once fed by the rainwater that percolated through the soil of the forest cease to flow. Moreover, rivers dwindle to a trickle in dry seasons and become destructive, raging torrents that overrun their banks when it rains.

Modern-day deforestation and overgrazing are producing a world-wide pattern of topsoil destruction similar to that which centuries ago occurred in most of the Moditerranean countries. Previously wooded, steep mountain slopes have been reduced to bare bedrock in large areas of Greece. Turkey, North Africa. Spain, Italy, and Yugoslavia. The burgeoning population of Nepal, another example, has been forcing hungry peasants farther and farther up the slopes of the Himalayas in search of cropland. As firewood has become increasingly hard to find in Nepal, even newly planted trees are being used for fuel, as are grasses and other hillside plants. Destructive landslides, as well as the silting and flooding of lowlands, have been increasing as a result. Thirty-eight per cent of the most densely populated areas of Nepal now consists of abandoned farmland that can no longer grow any crops—an astounding statistic. As wood supplies have declined, people have also begun a more extensive use of cattle dung for fuel. Although always used as fuel to some extent, dung formetry was used mostly to fertilize farm soil and to improve its water- and nutrient-holding capability.

Similar patterns of soil destruction exist on mountainous slopes elsewhere in the world. In South America, the densest populations have always been in the highlands and plateaus of the Andes. It has always been possible to farm the steep Andean slopes as long as strict rules are observed. Besides being terraced to control the runoff of rainwater, the land must be allowed to remain follow —that is, return to its natural condition—for 8 to 12 years or more between crops. Only this extraordinary precaution enables the Andean soil to absorb enough water and its organic content to increase sufficiently so that food crops can be grown on it for a year or two before it is again allowed to lie fallow. When this practice is not observed, the soil's structure deteriorates, and then water erosion takes a heavy toll. In the hills of Peru and Colombia, for example, major landslides have become more frequent, floods have worsened, and massive amounts of silt have been carried down into the valleys. As silt fills the rivers, they overflow their banks or change their courses, usually with destructive results. Several multimillion-dollar dams built to produce electric power have become useless within only a few years because their reservoirs are filled with silt.

Topsoil Depletion in the United States. It is widely believed that North Americans learned long ago to control soil erosion and that the problem no longer exists in Canada and the United States. Although the causes and means of controlling soil erosion are indeed known to scientists and government agencies, their well-established theories have proved difficult to put into practice.

Overgrazing. The allotment of more animals to a pasture than it can support results in overgrazing. Grass that is continually cropped very closely cannot survive dry periods, and eventually the soil becomes exposed to wind erosion. As a very general rule, it has proved best to remove no more than one-third of the edible productivity of a range in any one year. Such an approach actually improves, the range and its cattle during years of adequate rainfall. It also ensures a reserve of soil, root systems, and moisture that can save the range during drought years. Once extensive overgrazing has occurred in regions with low average rainfall, the transformation of fertile, arable land into desert results. Over 10 percent of the world's deserts have been created by human abuse of crop-and rangeland.

Wind and Water Erosion of Farm Soil. Loss of soil moisture can also be a serious problem for farms in grassland regions; moisture evaporates rapidly from bare soil, especially after it has been tilled. Harvesting of crops always removes soil moisture. Only if native prairie grasses or other soil-building plants are allowed to grow during occasional fallow years will soil moisture be restored.

Even the best-managed farms, those utilizing windbreaks of trees, contour cultivation, terracing, and winter crops to hold snow and retard wind erosion lose at least 1 per cent of their topsoil each year. Thus, even a well-managed farm that is kept in continuous cultivation can be ruined eventually; how soon this happens depends largely on the original depth of the natural topsoil.

Many United States farms are badly treated, despite such terrible experiences as the dust bowl of the 1930s. At the close of the 1970s, many agricultural and economic experts in the United States, responding to soaring wheat prices and a dwindling world food reserve, actually called for fence-to-fence planting of wheat. While this practice may indeed feed the world's hungry for a few years and even help control economic inflation, it will eventually produce even greater disasters. Resulting dust bowls could turn the Great Plains into a barren desert unsuitable even for light grazing.

The greatest threat to topsoil in the United States at present is water erosion in the corn belt. Most severely affected are southern Iowa, northern Missouri, western Tennessee, western Texas, and the Mississippi basin. Most of the lost topsoil is entering the Mississippi River and being carried into the Gulf of Mexico at an average rate of 15 million tons a minute!

By 1977, about one-third of all land in the United States

was being eroded at a rate that was noticeably reducing its productivity. In 1980 and 1981, an estimated 19.2 million hectares (48 million acres)—10 per cent of the total land area—were losing almost 6 tons of topsoil per hectare per year, each year's loss representing 9 years' accumulation under the best possible conditions. Much of this erosion is controllable and need not occur.

THE DUST-BOWL DISASTER OF THE 1930s

Large numbers of settlers began to arrive in the Great Plains in the 1880s. Each homesteader was deeded 64 hectares (160 acres) by the United States government, an amount judged to be more than enough for a family farm; 64 hectares was fully adequate in the eastern United States, but than area had a much higher annual rainfall than the Great Plains. The settlers plowed up the thick turf, with its long grasses and extensive root systems, and replaced it with corn and short rooted cereal grains, reserving some of the land for pasture. Harvests were bountiful, and the fature seemed bright. But the 1890s brought the first drought, and many settlers were forced to sell out and move away. In time, however, the rains returned and hopes rose once more. Then, the year 1910 brought another drought and with it the first huge dark clouds of dust. The plains states have the highest prevailing wind velocity of any part of the country. The winds tear at exposed, dry soil, lifting the fine silt particles high into the air, tumbling the larger particles into smothering drifts. More farmers left in 1910 and 1911, but many were encouraged to stay and even expand their holdings by two important developments: the advent of gasoline-powered tractors and harvest combines, and the soaring wheat prices caused by the disruption of European farming during World War I.

Drought, Dust and Disaster

The 1920s were years of increasingly heavy investment in raising wheat and cattle. Ranches and farm ranges were extended farther and farther into arid lands. The next drought began in 1931; it was to last through 1934. By the summer of 1933, millions of hectares of brown, shrivelled plants stood unharvested. Water holes dried up, cattle died, and thousands of people began a sad exodus to California, a tragedy made notorious by John Steinbeck's classic novel The Grapes of Wrath.

The spring of 1934 brought yet another crop failure and the first major dust storm. On April 14, large regions of Kansas and Colorado acquired an apocalyptic atmosphere as total darkness descended in the middle of the day. Dead birds and rabbits littered the fields. A month later, on the single day of May 11, a violent windstorm swept an estimated 350 million tons of rich organic silt high into the upper atmosphere, where it was caught by the high winds of the jet stream. New York City's skies darkened, and dust settled on ships hundreds of miles at sea. On that day Chicago received a fallout of 12 million tons of dust, some 4 tons per person!

Attempts at Conservation

Out of the dust-bowl disaster was born, in 1953, the United States Soil Conservation Service. Without question, this agency has done much to retard soil erosion and, in some areas, has even reversed the trend. Millions of hectares of land most vulnerable to erosion have been returned to pasture. Practices such as following; strip-cropping, the practice of alternating rows of cash crops with rows of sod-forming crops, such as hay: and leaving stubble and other crop residues behind at harvest to hold moisture and retard wind erosion during winter have been promoted. Terracing and contour cultivation of slopes and the planting of windbreaks of trees between fields have reduced wind erosion. Ranchers have been encouraged to limit herd densities to reduce over-grazing.

More Trouble in the Future?

Drought hit the plains again in the 1950s; and, although dust-bowl conditions did not develop, damage was severe. However, rains soon returned, and reasonably good yields were obtained again. But a new dust bowl may be in the making. As reserves of wheat to feed a hungry world have dwindled and prices have risen, the pressure to plant wheat has increased. In 1973 and 1974, 4 million hectares of pastures, woodlands, and idle fields were plowed and planted for the first time. Over half of these lands have been badly treated, and their annual loss of topsoil has averaged 27 tons per hectare. No more than 12 tons per hectare per year is regarded by government soil experts as a tolerable loss for cultivated land. In the Southern Great Plains, the 1974 losses on 27,000 hectares of newly planted land ranged from 34 to 314 tons per hectare! In the early 1980s, as grain prices rose and beef prices fell, strip-cropping and following were being eliminated, and dry range-land was being planted to wheat. The cutbacks that began in the mid-1980s were related to economic conditions rather than being based on environmental concerns.

Water Supplies

The supply of fresh water upon which most terrestrial life depends is provided by the hydrologic cycle. Evaporated water soon condenses and falls to earth as rain, sleet, or snow. One-fourth of it falls on land. However, precipitated water is but 0.0007 per cent of all the water on earth. When precipitated water seeps into the ground, it enters vast aquifers (from the Latin aqua, "water," and ferre, "to carry"), underground systems of porous, water-bearing rock. The flow of water in an aquifer can be as slow as a few inches per year. A continent's aquifers contain about 30 times as much water as all of its lakes, rivers, and streams. Eventually, precipitated water makes its way into the lower levels of aquifers and from there into the springs, rivers, and lakes. The existence of these sources of fresh water also depend on the hydrologic cycle.

Meeting Water Demand. The amount of the world's water supply is fixed; we cannot control the amount of percipitation, although to some extent we can channel the water once it falls to earth. The total amount of precipitation and its distribution throughout the year differ greatly around the world. These two factors are the key to the adequacy of the water supply for any region. The world's water supply has been calculated as 13.63×10^{20} l (3.59×10^{20} gallons) a day. Of this amount, the United States receives a sizable share, some 16.24×10^{12} l (4.3×10^{12} gallons) a day in rainfall and another 24.59×10^{11} l (6.5×10^{11} gallons) a day available from lakes and rivers. This is about 50 times the country's average yearly demand and thus, in total, would seem to be far more than adequate. Yet water shortages occasionally occur in several parts of the country. One reason is that some high population densities have developed in regions with very low rainfall, necessitating the piping in of water from great distances. Another is that irrigation, which typically either uses more well water than is replaced by precipitation or uses water brought from distant sources, forms the basis of agriculture in several regions.

Wastefulness Versus Conservation. A major reason for shortages in densely populated areas is that people use highly purified drinking water inappropriately—for such needs as washing cars, floors, side-walks, and dogs; watering lawns; and the like. The average United States household uses about 680 l (180 gallons) a day! Thus, most threatened water shortages in the United States are more accurately threats to a tradition of wastefulness. In an experiment, waste water from bathing and laundry was stored and used for flushing toilets; total consumption was reduced by 39 per cent. Traditionally, industry has also used far more water than necessary. This is demonstrated by the fact that when the Kaiser Steel Company of California decided to recycle water, it cut its use from 246,000 l (65,000 gallons) per ton of steel to some 5,300 l (1,400 gallons) per ton.

Worldwide, total water use is actually less than a third of the amount annually available. But demand is increasing and is expected to exceed 50 per cent of availability by the year 2000. It is thus obvious that more stringent conservation efforts are necessary. In addition, the cost of piping massive amounts of water from distant sources will eventually exceed what the public is able or willing to pay. This may impose local limits on population expansion in some areas. Los Angles, California, for example, has an annual rainfall of 38 cm (15 inches), enough to support only a small city. In 1920, an aqueduct 483 km (300 miles) long was built, bringing water to the city from Owens Canyon, high in the Sierra Nevada range. In the 1930s, increased demand for water led to construction of a second aqueduct, 322 km (200 miles) long, from the Colorado River to Los Angeles. In the 1960s, the Feather River, a tributary of the Sacramento River, which was already supplying water to San Francisco, was tapped to meet Los Angeles' ever-growing demand for water. Diverting water from its normal course has been destructive of natural ecosystems and has severely limited crop productivity in regions such as Owens Canyon. Each of these developments was marked by bitter legal battles fought by the angry residents of the areas deprived of water by the projects.

City officials in Los Angeles are now seeking additional sources of water from the low-population wilderness areas in the northernmost part of the state. They hope to divert the Eel and Klamath rivers, the last free-flowing rivers in the state, into the Sacramento River system to provide additional water to both the Los Angeles and the San Francisco areas. Rivers are usually diverted by damming them. Dams have their own environmental impact.

Problems Accompanying Irrigation

Irrigation is an ancient practice, probably dating from the time that humans first switched from hunter-gatherer cultures to ones based on agriculture. In lands where rainfall is sparse or irregular, crops need watering if they are to thrive. Two sources are available; surface waters, consisting of lakes and rivers; and groundwater, or well water. With either source, problems have usually accompanied the prolonged practice of irrigation.

Soil Salinization. Eventually, salinization of the soil occurs in any irrigated area that is warm and dry and is poorly drained. Irrigated farms downstream from the Aswan High Dam are being ruined by salt residues, and many are no longer cultivated. Many of the fertile farms of California's hot and arid Imperial Valley have also been abandoned because of salinization. Irrigation water from the Colorado River, which is not unusually salty, deposits 3 million tons of salt in the valley's soil each year. The problem has been attacked in many parts of the valley by the construction of miles of costly drainage ditches into which the accumulated salts are flushed periodically. This salt-laden water is returned to the Colorado River, increasing its salinity by 30 per cent in the last 20 years. In response to complaints of Mexican farmers, much of this water in now being desalinized at considerable expense before being returned to the river. The salt is being discharged into the Bay of California.

Depletion of Groundwater. Many areas of the Great Plains in the United States obtain up to 75 per cent of their water from wells. For many years, the amount of water drawn from wells had no appreciable impact on the level of the water table, the depth below which the ground is saturated with water. However, with the expansion of irrigation argriculture since the 1950s, this has changed. In southern Texas, which by the 1980s was growing 25 per cent of the cotton produced in the United States, the number of water wells multiplied 15-fold, from 2,000 in 1946 to 30,000 in 1966. During that period, the water table dropped 120 m (400 feet) in regions of Arizona that used well water for extensive irrigation. As a result, some 128,000 hectares (540 square miles of cropland had to be abandoned.

The high plains of western Texas and Arizona draw upon the Ogallala Aquifer for their water. This mammoth underground supply represeats water that has underlain the southwestern United States since the putting of the last of the great glaciers to cover much of the country, thousands of years ago. Because it is not replenished by precipitation, this water supply, like deposits of fossil fuel, must be regarded as a nonrenewable resource.

California's San Joaquin Valley also makes extensive use of well water to maintain its high productivity; by the early 1980s, it was producing an annual overdraft of 246 million 1 (65 million gallons) Many years of overdrafts have resulted in serious soil subsidence, or settling, in the San Joaquin Valley; some areas have sunk by as much as 10 m (33 feet), but other examples can be found elsewhere in California as well. Such extensive subsidence compacts the soil, rendering it incapable of absorbing the amount of water it originally contained even if irrigation were halted.

On a nationwide basis, the United States withdraws twice as much water from its aquifers each year as is restored by the hydrologic cycle. Land stosidence has been particularly destructive to some residential and industrial areas. Some homes near Houston, Texas, for example, have such to a point at which they are often flooded by ocean tides. Moreover, coastal areas are experiencing extensive saltwater intrusion into the land, which ruins its wells and destroys its capability of supporting almost all types of life long adapted to it.

The wisdom of irrigation is now being seriously questioned. Farms are using water at an unprecedented rate, four times the amount used by cities. Although the water used in irrigation is only 60 per cent that used in industry, most industrial water can be recycled; irrigation water, however, is lost. This makes growing food by irrigation methods very expensive. According to one estimate, it requires 10,000 to 50,000 tons of water to raise each ton of food produced by an irrigated farm in an acid region.

Destruction of Forests

Few human depredations can match what has been done and is still being done to the world's forests. In the nineteenth century, when there were no federal or state agencies overseeing

*An overdraft is the amount of water removed in excess of that returned by precipitation.

the use of public lands, it was common for small American lumber companies to obtain lumber rights over a region, clear-cut all its marketable trees, and leave behind a denuded area whose topsoil was soon eroded by rains. The slash—consisting of cut-away branches and foliage—left behind after the logs were dragged out soon became tinder dry and was easily ignited into roaring fires by lightning. The fires that resulted from ignition of dried slash between 1850 and 1900 were of an unprecedented magnitude. The great Peshtigo, Wisconsin, fire in October 1871, which was fed by dried slash, burned over 400,000 hectares (1 million acres) of forest and claimed between 1,200 and 1,500 lives. In Wisconsin alone, there were 2,500 fires from 1880 to 1890, burning an average of 200,000 hectares (half a million acres) each year.

Modern Conservation in the United States. Government agencies, such as the United States Forest Service, and private citizens' groups, such as the Sierra Club, now exert moderately effective control over American forests. One practice aimed at preventing soil erosion is the selective cutting of only some of the marketable trees from any one stand. Those allowed to remain provide the seeds for new growth. Such a system ensures a sustained yield, the forests being cut selectively about every 10 years. The technique is expensive, however, and works well only with hardwoods, Such as beech and maple, whose seedlings can grow in shade.

Another method of forest conservation, called monoculture, involves clear-cutting stands of trees of about the same age and then reseeding or replanting the area with one or a few species. This tree-farming technique permitss harvesting such giants as the Douglas fir, whose selective cutting is impractical because of the damage caused to other trees when the firs are felled. Moreover, their seedlings do not grow in shade. Clear-cutting is controversial, however, in part because of the erosion and loss of soil nutrients that invariably follow when trees are cut from slopes. Monoculture also makes trees more susceptible to disease and insect pest epidemics than are natural populations,.

in which individuals of the same species and age are widely dispersed.

Despite all regulations, the United States still cuts down more trees than it regrows. Current projections are that the country will experience a 265m³ million (9.3-billion-cubic-foot) annual deficit by 2010, with demand exceeding supply by 50 per cent. Forest conservation needs to be intensified, not only to meet demands for wood and wood products, but because forests constitute an important part of the world's ecosystems.

Worldwide Destruction of Forests. Close to half of all the world's trees have been cut down since 1950. Two-thirds of Latin America's forests and half of Africa's have been depleted. During the 1970s alone, Thailand lost one-fourth of its trees, and the Philippines one-seventh. One reason for the destruction is slash-and-burn agriculture, an ancient practice which has been increasing as world populations have grown. It Involves cutting down all the trees in a chosen area, burning them, and raising crops for a year or two in the ash-enriched soil. Overall, there is a net loss of beetween 1 and 2 per cent of the world's forests each year. This rate is greater than could be compensated for by even the best efforts at reforestation.

A growing need for firewood, which for 75 per cent of the world is the only fuel used for heating and cooking, is one reason for so much destruction. For example, over 90 per cent of all wood cut in Africa is burned for fuel. The worldwide shortage of firewood is a far greater energy crisis than the soil shortage. By the year 2000, firewood needs are expected to exceed supply by 25 per cent. Where firewood shortages have developed, the substitution of dried animal manure as fuel has reduced African food production by 20 million tons of grain a year. In Nepal, the floods caused by denuding slopes of their trees for use as firewood cost millions of dollars in damage and and kill thousands of Nepalese and Indians each year.

Wildernesses as Natural Resources

A wilderness is best defined as an extensive, compact tract of land into which human intrusion by roads, motorized vehicles, prospecting, mining, lumbering, management, or control does not occur. A true wilderness forest must be free to burn and regrow, even to die from diseases. Fallen dead trees must be left to rot, rivers to flood or change their courses, and moose and deer to starve to death when caught in deep snow. These are all natural processes to which wilderness communities have become adapted without human assistance in the course of the evolution of their various members.

Why preserve such extensive tracts of land, especially ones that could serve as sources of greatly needed wood, oil, or valuable minerals? Scenic beauty is one reason that is often given as the answer. Yet, as important as is the natural beauty of a wilderness for its recreational value and its capacity to gratify our aesthetic sense, these functions could probably be served by far less than the many millions of hectares now officially preserved as wildernesses. The preservation of extensive wilderness areas is much more crucial because of their role in ensuring the survival and variety of species, a point briefly mentioned in Chapter 35.

The Need for Extensive Tracts as Species Reservoirs. One of the important functions served by wildernesses is as reservoirs of many species that would otherwise become extinct. Disturbing a wilderness or reducing its size below a certain level is bound to reduce the chances of many of its species for survival. For example, to support a family of four grizzly bears in Yellowstone National Park requires an average of 300 km^2 (115 mi^2) of woodland. Yet, of the 89 officially protected wilderness areas in the western United States in the late 1970s, only 10 were larger than 1,000 km^2 (386 mi^2).

The territory sizes required by birds differ greatly from one species to another. In one study, eastern robins were found to need only 1,200 m^2, and red-winged blackbirds only 3,000 m^2. The great horned owl was found to require 50 hectares (500,000 m^2), and a family of golden eagles a territory of up 9,300 hectares. As the quality of an environment deteriorates, the size of the territory required by a given species usually increases; robins in a city need more area than do robins in the countryside.

The examples given here are of large and conspicuous species whose requirements have been discovered by painstaking ecological investigations. Wildernesses contain many species whose essential requirements have never been determine and, indeed, many species that have not yet been discovered, named, and described.* Tropical forests in particular are valuable species reservoirs. Although they cover only 10 per cent of the earth, they harbor nearly half of all species. If present practices of cutting down rain forests continue, some 20 per cent of these species are expected to become extinct within the next 20 years. Thus, the problem of survival affects more than the few dozen well-known endangered species, such as the redwoods, whooping cranes, elephants, and whales; it concerns hundreds of thousands of unique and irreplaceable forms of lise that will have vanished from the earth forever.

The Need to Preserve a Great Variety of Species. Ecosystems require the presence of many species for several important reasons. In a complex natural ecosystem, each type of organism has its own approach to processing the forms of energy and materials upon which it subsists. Various species of insects, worms, fungi, protists, bacteria, and even the symbionts inhabiting the intestines of cockroaches and termites each play a role in the constant recycling of the materials present in the environment.

Purifiers of Polluted Air and Water. Many of the materials found in nature, including many produced by organisms, are toxic for some kinds of orgonisms. Yet these same materials, many of them toxic for humans and domestic animals, are

*Best estimates place the total number of species on earth between 3 and 10 million, with onlv 15 to 20 per cent having thus far been named and described by scientists.

metabolized and degraded to simpler, nontoxic forms by various other organisms. Only complex communities contain sufficient diversity of species to accomplish these detoxification processes efficiently.

Wildernesses are highly effective in metabolizing the many poisonous substances with which we pollute our air and water, returning the air and water to us in purified condition. Despite its large size, New York's Central Park does not qualify as a wilderness. Yet it serves a similar purpose to a limited degree. For example, the air's sulfur dioxide (SO_2) concentration over the park is measurably below that of adjacent areas. When a breeze blows away any city's accumulation of smog or when a rain washes its air of noxious fumes, these materials are carried to some other part of the planet, where they are usually degraded or otherwise detoxified. Wildernesses play a large role in this valuable process.

Vital Needs in Biological Research and Medicine. A second important reason for preserving as many species as possible is that so many of them are useful in biological research or in medicine. Some species of little or no commercial value have often proved to be the material of choice for the investigation and discovery of basic biological mechanisms. Fruit flies, bread mold, and sewage bacteria are the materials on which several Nobel-prize—winning investigations of genetic mechanisms were conducted. Important breakthroughs in our understanding of nervous function have resulted from studies of certain squids, sea slugs, and jellyfish. Our knowledge of embryonic development has been advanced through investigations of African frogs, salamanders, and sea urchins. A species of coral was recently found to be an excellent source of precursor molecules used to synthesize prostaglandins, a valuable drug.

How Wilderness Species Benefit Other Ecosystems. The many species in our wildernesses constitute a reservoir of adaptable organisms capable of recolonizing damaged and devastated areas. For example, supplies of readily available species capable of adapting to a particular situation have sometimes restored areas to more natural conditions; again, New York City's Central Park is a case in point. The species of plants, birds, and small mammals that recolonized this area, which in 1850 had been virtually stripped of its trees and shrubs, came mostly from wilderness areas. Even if all of the world's species were known to us, only a small fraction of them could ever be conserved in botanical gardens and zoos. Moreover, only a few animal species are able to reproduce in captivity, and, of those that do adapt to life in a cage, few would survive if returned to the wild. As our wildernesses deteriorate, the ability of other ecosystems to recover from damage or even to maintain their own viability is being seriously jeopardized.

The Need for Concern. Some of those who believe that efforts to halt the extinction of species are misplaced have pointed out that, over millions of years, far more species have become extinct through natural causes than now exist on earth. What harm, it is said, is there in the extinction of a few more? The answer lies in the considerable difference between these two modes of extinction. Most species that become extinct naturally do so in a slow, gradual process over centuries and millenia. This permits evolutionary adaptations of other species to the altered situation. Indeed, the adaptation of competing species is a major cause of such extinctions. However, the changes now being wrought are unprecedented. Between 1600 and 1900, know mammals and bird species were eradicated at a rate of one speciesevery 4 years. During the last 80 years, the rate increased to one speicies every year. However, these were only the known species; the actual number of fish, invertebrate, and plant species now undergoing extinction cannot be determined.

It is a well-established principle that perturbations of one parameter in a natural ecosystem not only can be felt throuhout the ecosystem but often become amplified. Many of the species of a large ecosystem have undergone a long coevolution culminating in a beautifully harmonious adaptation to each other. Extensive destruction of a large ecosytem of extermination of many of its species upsets this balance, leading to overproduction of some species and underproduction of others. The ultimate effects upon surviving species are unkown.

Another reaction to the movement to preserve wildernesses is the suggestion that some wildernesses are of little or no value and ought not to be protected. The ones usually mentioned are deserts, marshes, swamps, and tundras. Again, we do not know what effects—desirable or undesirable—would result from extensive losses of these ecosystems.

DIRECT DESTRUCTION OF NATURAL ECOSYSTEMS

Many commercial and public works projects have substantially altered and even destroyed unique natural ecosystems. Many of these systems are—or were—the homes or breeding grounds of uniques species. Mining the elimination of wetlands, and flood control measures have all resulted in the direct destruction of natural ecosystms.

Mining. Besides contributing to a runoff of toxic substances, the tailings left by mining operations represent a destruction of both the natural role and the beauty of an area. Strip mining is destructive because it removes whatever type of habitat is present. However, some parts of the United States damaged by strip mining are being reforested, and ponds created by the mining are being stocked with game fish.

Draining and Filling in Wetlands. Among the most harmful projects from the standpoint of adverse environmental impact are the draining and filling in of marshes, shallow bays, and other wetland areas. Besides destroying wildlife that uses a coastal bay the year-round, dredging and filling ruin its usefulness as a seasonal spawning ground for many marine animals and as a stopping place for migratory waterfowl. Three rather well-known examples of such landfill operations are those of California's San Francisco Bay and Florida's

Tampa and Biscayne bays. Tampa Bay is more than 20 per cent filled. By the middle 1960s, 60 per cent of San Francisco Bay had been filled.

The fill used in such projects is usually obtained by dredging other areas of the same bay, a process in itself detructive of the valuable wildlife habitats the bay provides. All along both United States seaboards, the draining, dredging and filling of wetlands has been used as a way of creating commercially valuable land. A 1980 study reported that 40 per cent of the coastal wetlands of the 48 contiguous states has now been destroyed. The effect of such massive removal of ecosystems upon the rich fauna and flora of the Atlandtic and Pacific coastal waters is thought to be enormous.

The Effects of Flood Control on One Ecosystem. Flood control and swamp drainage projects in central and southern Florida have had adverse ecological effects. The southernmost 160 km (100 miles) of the Florida peninsula, south of Lake Okeechobee, constitutes a unique subtropical ecosystem of great complexity: the Everglades. This region enjoys a moderately warm climate and a heavy annual summer rainfall that until early in this century produced a substantial overflow of Lake Okeechobee, This runoff moved slowly southward in a broad shallow sheet that covered much of the land immediately south of the lake. The marsh ecosystem supported by this abundant water was marked by numerous small is lands, occasional sloughs (marshy inlets) and "gator holes" (deep depressions in which alligators live), and extensive dense growths of tall saw grass rising from the muck soil of the marshes. At the peninsula's tip, where the flow of fresh water meets and blends with seawater, the saw grass marsh gives way to heavy stands of mangrove trees. The brackish waters of the mangrove swamps provide spawning grounds for a great variety of fish and invertebrates. To the southwest of Lake Okeechobee is the extensive Big Cypress Swamp, its dark waters supporting another complex community of organisms.

The dominant animals of this ancient ecosystem were the

alligator and the panther. Although poaching is restricted, alligator populations continue declining and the panther is now seldom seen. Eighty years of "development" of Florida have rendered this highly mature community very unstable; its disintegration is well under way.

Records of engineers examining the commercial potential of the Everglades about 1900 described them as beautiful and awesome but also as "entirely worthless to civilized man in its present condition." In 1906, construction was begun on a canal to drain Lake Okeechobee into the Atlantic. This project was the start of what was to become a vast network of canals, levees, dikes, and spillways. The purpose of this project was to drain the marshy soil for agricltural use and residential development. The system subsequently has proved inadequate to handle the runoff during wet years, yet it overdrains the land during drier years. During droughts that hitherto had little effect on the glades because of the flow water from the north, the soil dries out. Normally, the peat muck soil* of the glades acted as a giant sponge that soaked up water during rainy periods and slowly released it during dry periods. However, diverting Lake Okeechobee water to the ocean during dry periods has resulted in dried, cracked soil; the death of natural communities; and crop failures. Attempts to regulate the flow to downstate regions have often proved unsatisfactory. The adverse effects include the following:

— Peat muck fires: before the drainage canals were built, saw grass fires* were common during the dry season; on occasion, the exposed peat muck also caught fire and smoldered for a brief time. Since the land was drained, however, peat fires often burn uncontrollably for months at a time. Fort Lauderdale and Miami have many dark

*Peat muck is a spongy organic soil so rich in partially decayed vegetation as to be combustible when dry.

**Occasional fires are one of the factors that maintain a grassland community, discouraging growth of trees.

days because of muck fire smoke. The fires destroy much natural wildlife and valuable agricultural soil.

— Soil subsidence: once muck soil is exposed to air, it undergoes, both oxidation and drying, shrinking and compacting as a result. Although muck soil requires about 400 years to form to a depth of 33 cm (1 foot), it has been disappearing at the rate of 33 cm every 10 years.

— Depletion of "gator holes": at one time, there were an estimated 1 million alligators in southern Florida. Many lived in "gator holes," which also provide a habital for many unique species of fishes, birds, and invertebrates. Few of these communities remain.

— Saltwater intrusion: because the Everglades are barely higher than sea level, the fresh water that normally flowed down from Okeechobee prevented the intrusion of seawater into the soil of southern Florida. After initiation of drainage, salt water began to seep into the freshwater aquifers nearest the sea coast, resulting in contamination of wells with salt and salinization of the soil, which is destructive of natural terrestrial communities and harmful to crops. Natural aquatic communities are also being destroyed as seawater intrudes into the fresh water of brackish swamps and marshes.

POLLUTION

Environmental pollution can be defined as the direct or indirect impairment of the environment's suitability for supporting life by harmful concentrations of materials, whether or not the materials are toxic. Thus, a very-high concentration of a nutrient might be polluting, whereas a deadly poison in great dilution is not.

Earlier in this century, wastes could be discharged into the environment with relative impunity. Two factors have changed this. One is the population explosion; more people mean more pollution. The other is the rapid advance of technology and affluence that has occurred in the industrialized nations and which has increased the production of toxic pollutants.

In the course of evolution, the wastes produced by one kind of organism eventually became the life-sustaining nutrients of various other organisms. Modern human activity, however, has altered this relationship. As world populations have increasingly become concentrated in cities*, the natural wastes they produce have contributed significantly to environmental pollution. Yet natural wastes are not the most harmful pollutants being generated.

Unlike biological wastes, the wastes that result from mining, processing of raw materials, and manufacturing are often highly toxic. Even when they are biodegradable, that is, can be degraded by natural decomposers in the soil, rivers, lakes, and oceans, some wastes are released into the environment in concentrations and volumes that overwhelm the capacity of the decomposers to degrade them, thereby threatening the existence of many aquatic organisms.

Liquid wastes that are both nonbiodegradable and toxic to one or more forms of life pose more persistent problems. If released into waterways, they kill or prevent the reproduction of much aquatic life and render the water unfit as a source of food organisms and for drinking and recreation. If stored on land, they may contaminate groundwater and thus poison springs, wells, and spring-fed lakes.

Some specific examples of environmental pollution, some of their known ecological effects, and a few approaches to solving the problems they create are briefly examined in the following sections.

*Early in this centuty, London became the first city to reach a population of 5 million. By the year 2000, there will be 60 cities with populations of over 5 million, some of them 20 or even 30 million in size. Over half of the world's population will then be living in cities.

Air Pollution

Because of air's mobility, it has a high natural capability for reducing the impact of air pollution on ecosystems. However, when winds are calm in areas of high population density, air pollution can be extremely serious and even life-threatening. For example, during one brief period in 1952. smoke density and sulfur dioxide (SO_2) concentrations in London rose above normal, and death rates soared from about 250 per day to a peak of nearly 1,000. In 6 days, 4,000 more people had died than was normal for the period, and during the next 2 months, 8,000 more people than usual died.

Most people can tolerate some air pollution for extended periods of time. However, long-term exposure to air pollutants has many adverse effects on health. The largest volume and the most serious air pollutants, many of which are carcinogenic, are emissions from the burning of fossil fuels. Moreover, when various hydrocarbons and nitrogen oxides from automobile emissions combine under the action of sunlight, they form ozons (O_3) and the other health-threatening constituents of the atmospheric haze known as photochemical smog.

One of the areas most notorious for smog episodes is Los Angeles, California. whose many automobiles and abundant sunshine produce substantial amounts of smog. The problem is serious because of the city's frequent temperature inversions, which result when a cool, humid breeze flows in under a stationary high-pressure layer of warm. dry air (ordinarily, air temperature decreases with altitude). Pollutants generated at ground level become trapped under the impenetrable layer of warm air. The mountains that form a broad semicircle to the north of the Los Angeles area enhance the polluting effect by occasionally trapping the inversion in the basin for many days at a time, with air quality worsening daily.

Temperature inversions and health-threatening air quality crises also occur in the midwestern United States whenever highs (high-pressure weather systems) stall for several days during warm weather. A major episode occurred in August 1969 that affected several large cities for up to 10 days. The crisis is listed as episode 104 in United States Weather Bureau records because it was the one hundred and fourth high-air-potential (HAPP) episode that covered more than 194,000 km² pollution (75,000 square miles) and lasted more than 36 hours since monitoring began in the early 1960s. HAPP episodes covering less area are not numbered; they occur more than 25 per cent of the time over almost all of the United States.

Long-Term Effects of Moderate Air Pollution on Health. Long-term, moderate air pollution actually causes more disease and death than do acute crises. Long-term exposure to air pollutants increases the risks of contracting cancer and of developing respiratory diseases such as emphysema. Although precise data about the long-term effects on health of each of the various pollutants are difficult to obtain, it is well agreed that air pollution is harmful to health. One somewhat useful approach to studying its effects has been to compare mortality (deaths) and morbidity (serious illness) data for a given city with the concentrations of various pollutants found in its air. For example, there is a high correlation between the concentration of heavy industries and the number of deaths due to cancer. World Health Organization data indicate that 80 to 90 per cent of all human cancers are environmentally related or induced. Certain types of cancers occur in greatest frequency near factories producing certain products, for example, liver cancer incidence is high near synthetic rubber factories. Air pollution is the suspected cause.

Our understanding of the injurious effects of long-term exposure to some of the airborne pollutants in cities has been enhanced by studies of cigarette smokers. Cigarette smoke contains many of the same pollutants found in the air of large cities, for example, carbon dioxide (CO_2), carbon monoxide (CO). and at least seven polycyclic (multiple ring) hydrocarbons that can produce cancer in animals under certain conditions. Numerous studies have confirmed that the incidence of lung cancer, emphysema, and coronary artery disease is far greater in smokers, whether they live in rural cities or in environments, and is in direct proportion to the number of cigarettes they smoke per day and the number of years they have smoked.

Acid Rain. An increasingly serious problem has lately resulted from fossil fuel combustion products, the most troublesome of which are nitrogen oxide (NO and NO_2), hydrogen sulphide (H_2S), and sulphur dioxide (SO_2). Dissolved in atmospheric moisture, these compounds produce nitric acid (HNO_3) and sulphuric acid (H_2SO_4), both strong mineral acids. The resulting acid rain has destructive effects on freshwater and soil ecosystems, including ones thousands of miles from the pollution source.

The Control of Air Pollution. There are two basic, practicable approaches to controlling the quality of the air we breathe. One is to limit the pollutants emitted into the air, for example, by the burning of sulphur-free oil and coal, or by using sophisticated, expensive technology that is still being developed. The other approach is to choose alternative technologies or sources of energy. Wind, water, and solar power, for example, are far less polluting than is the burning of any fossil fuels. Likewise, the use of bicycles and battery-powered cars is far less polluting than the use of vehicles powered by internal combustion engines.

WATER POLLUTION

Water pollution is usually defined as the addition of materials to water in such quantity as to lessen its suitability for the life of aquatic organisms, for irrigation, for recreation, or for drinking. Some water pollutants are inherently toxic to one or more forms of life. While seldom intrinsically toxic, nutrients may be toxic in high concentrations or may produce so much growth of bacteria or other aquatic life as to make life impossible for other aquatic organisms.

Sources of water pollution include human wastes; runoffs of

industrial processes, farmlands, feedlots, and mines; air pollutants that find their way into lakes and rivers; accidental and deliberate discharges of petroleum; and radioactive wastes. Farm runoffs include fertilizer, manure, insecticides, and herbicides.

The volume of water pollutants in the United States is huge, Runoffs into the Ohio River from mines in Pennsylvania and West Virginia, for example, account for a daily outpouring of 200,000 tons of sulphuric acid. The Detroit River alone dumps 20 million tons of miscellaneous waste into Lake Erie every day. Cleveland, Buffalo, and Toledo all add their sewage effluent and industrial discharges to the same lake. By 1980, 11 regions along the shores of the Great Lakes alone had been designated areas of major pollutions.

Several major water pollutants are considered in the sections that follow.

Biodegradable, Nontoxic Organic Wastes. The ways that sewage is usually treated in American cities tells something of the nature of the problem of pollution by biodegradable, nontoxic organic wastes.

After secondary treatment has greatly reduced biological oxygen demand (BOD), this sewage effluent is allowed to enter a river, lake, or ocean. The BOD of water is defined as the amount of oxygen that is removed from the water by the respiration of sewage bacteria in the course of 5 days at 20°C as they decompose its organic matter; as such, it is an index of the water's content of biodegradable organic matter. Thus, the greater the bacterial growth in a water sample, the more oxygen is used, and the greater its BOD. A high BOD and consequent high depletion of oxygen will slow the process of sewage degradation. Secondary treatment usually reduce the BOD by 60 to 90 per cent. The remaining organic matter in the effluent then decomposes in the lake or river into which the effluent is discharged, usually without promoting enough bacterial growth or using enough of the water's oxygen to jeopardize its acquatic life.

BOD, however, is only one index of water quality. The bacteria that decompose the organic matter release inorganic nutrients, such as phosphates and nitrates, into the effluent. This promotes eutrophication of lakes and streams in which such nutrients have been the limiting factors in algal growth. Lakes, especially those with weak currents or still waters, may experience growths of algae massive enough to form a heavy surface scum that reduces the penetration of light. The excess algal growth dies and decomposes, a process that depletes the water's oxygen, thereby killing aerobic acquatic life. The scum and its odour limit the lake's usefulness for recreational purposes and as a community water supply. This process may initiate or accelerate ecological succession. Many communities now follow secondary treatment of sewage with tertiary treatment, a process that removes toxic metals and 90 per cent or more of the phosphates, nitrates, suspended solids, and bacteria.

The burden of biodegradable organic pollution of waterways in the United States has risen steadily in the last 25 years. Only about 15 per cent of this, however, can be attributed to population increase. The rest is due partly to increased use of disposable products that end up in substance present in great dilution in the environment may thus reach lethal concentration in the top consumers. For this reason the public is often warned against eating large freshwater fish caught in polluted waters.

Much of the toxic waste that pollutes water supplies does not enter lakes, rivers, or the sea. United States industries alone dispose of or store some 190×10^9 l (50×10^9 gallons) of liquid waste each year in dump sites or surface impoundments (pits, ponds, and so on). The United States Environmental Protection Agency (EPA) estimates that until the late 1970s, up to 90 per cent of these wastes had been disposed of improperly. By 1981, the EPA had registered over 52,000 hazardous waste dump sites in the United States. Over 12,000 of these were said to pose a "substantial and imminent threat to human health through contaminated groundwater, excessive radiation, or fire and explosion." Of 27,000 such impoundments inspected in 1980, 8,000 were found to be located on permeable soils above usable groundwater supplies. Over half of the groundwater thus far tested in these areas was already contaminated in excess of acceptable health standards. The seriousness of this problem stems from the fact that half of the country's drinking water is obtained from wells and springs supplied by groundwater. The United States Council on Environmental Quality reported in 1981 that well in 40 states had been shut down in the preceding 3 years because of contamination with toxic organic chemicals. By 1981, all three major aquifers on Long Island, New York which serve nearly 3 million people, were contaminated by organic chemicals, which resulted in the closing of 36 of the area's public water supplies.

Although sunlight and aquatic organisms may eventually purify surface waters of toxic wastes, aquifers lack these benefits. Once contaminated, an aquifer is likely to remain so for decades. The toxic waste problem, it is widely agreed, is the most serious of all environmental problems now faced by the United States. One of the last acts of the Ninety-sixth Congress in 1980 was to establish a superfund to clean up the worst of the dump sites. However, a 1976 act intended to regulate the disposal of hazardous wastes had not yet been fully implemented by 1985.

Thermal Pollution. Because warm water holds less oxygen and other soluble gases than does cold water, the heated water discharged into rivers and lakes by some industries makes life impossible for many kinds of fishes and other aerobic aquatic life. Such thermal pollution also prevents or delays a lake's springtime turnn over—the mixing of surface and lower layers; it thus retards normal planktonic growth. Although warm water itself does not kill certain species of fishes, some lethal fish diseases become more prevalent in warm water. Moreover, warm temperature is often the stimulus for breeding, thereby promoting reproduction at the wrong time of year. The eggs of fry of many fish species fail to develop in warm water. Warm water is a barrier to the spawning migrations of some fishes and can fatally mislead some fishes that normally migrate toward warmer waters at certain times of the year.

How much thermal pollution a particular lake or stream community can withstand depends on such factors as its volume, normal temperature, and importance as a species reservoir. Most of the harm from thermal pollution is a result of the fluctuating nature of such pollution. Whenever an industrial or power plant reduces or shuts down operations, water temperature suddenly drops, only to rise again when operations resume. Few members of warm or cold water communities can survive such shocks.

Oil Pollution. The phrase oil pollution usually evokes an image of a shipwreck, perhaps that of a supertanker, or the blowout of an offshore drilling operation, with hundreds of thousands or even milllons of liters of crude oil pouring into the ocean. Such accidental spills have indeed contributed significantly to the pollution of the world's oceans. However, despite the drama attending such disasters, they account for only 10 to 14 per cent of all the oil spilled into the ocean. Most of the rest comes from such practices as the deliberate dumping of waste oill, the discharging of oil form refineries, the flushing of ships' bilges, and much secret, illegal dumping. About 600,000 tons of oil a years also enter the sea as natural seepage from underground deposits. Additional sources of the hydrocarbons polluting the world's oceans are automobile and other internal combustion engines. Some 90 million tons of hydrocarbons enter the air each year from these two sources, and up to 50 per cent of that is believed to end up in the sea.

The effects of oil pollution are only beginning to be known. The damage easiest to determine is that done to fishes, birds, and seashore communities devastated by a crude oil spill. Besides immediate deaths of fishes and birds, the spawning grounds of oysters, fishes, and other commercially valuable species may be ruined for many years after a major oil spill in their vicinity.* Studies of the effects of oil pollution on the hundreds of thousands of other species that comprise marine communities have only begun. Many of the studies are directed towrad establishing a fund of information about the kinds and densities of species in various types of marine ecosystems to provide a basis of comparison for studying the effects of future oil spills. If industries and governments are to be made to bear the cost of oil pollution abatement, they must first be presented with the hard facts concerning what damage the pollution has done.

Except for our knowledge of the severe damage to coastal communities caused by major oil spills, we are largely ignorant of the toxic effects of hydrocarbons on the world's marine ecosystems. The greatest fear is that by the time the adverse effects of marine hydrocarbon pollution are finally understood, the pollution burden may have already passed a critical stage. Because hydrocarbons degrade very slowly, their effects would probably continue to be left for many years to come. The productivity of the sea is dwindling. Although this is no doubt partially due to overfishing, it is important that we know how much is also due to various types of pollution.

RADIOACTIVE POLLUTION

Environmental contamination by radioactivity is feared because of its mutagenic and lethal effects. The higher the dosage sustained or the longer it is sustained, the greater the biological damage. Radioactive damage to an ecosystem is not temporary, since radioactive pollutants may persist for thousands of years. Some forms of radioactivity become more concentrated as they are transferred up a food chain. Moreover, the

* Incidentally, the use of detergents to rid beaches of unsightly and noxious crude oil after a spill has proven much more harmful to marine organisms than the oil itself.

individual atoms are recycled repeatedly, with the potential for doing damage at every step along the way.

Operational Contamination of the Environment. One type of radioactive pollution is the contamination that arises from normal operations in the production of nuclear weapons and the operation of nuclear power plants. These operations include the mining, processing, transporting, the storing of radioactive ores; the normal operating of power plants and weapons arsenals; the storage and reprocessing of spent fuel; and the shortage of the radioactive components of decommissioned overage facilities. The possibilities for environmental contamination during the process of producing electricity from nuclear fuel can occur

1. during the mining and processing of ore,
2. during the operation of nuclear power plants,
3. during the reprocessing of spent fuel, and
4. in the storage of radioactive wastes.

Because the mine tailings of uranium ore extraction are themselves radioactive, they pose a threat to the health of miners and anyone else who lives or works in their vicinity for a time. The processing of uranium ore involves its enrichment to increase the proportion of uranium 235(^{235}U) by three to four times the natural level. Radioactive contamination of the environment generated by the enrichment process is minimal and not regarded as hazardous.

In the operation of nuclear power plants, minor defects in boiling-water nuclear reactors result in small leakages of radioactivity into the cooling water circulating through the reactor core and thus into the environment. Some krypton. 85 also escapes as a gas. Both of these "routine emissions," however, are at such ,a low level as to be comparable to the normal background of radioactivity from rocks and the cosmic rays from outer space. Barring accidents, nuclear power plant operation is not regarded by the industry or the United States government as a significant threat to health.

In a conventional reactor, after a year or more of a operation, fission products accumulate in the fuel rods in amounts sufficient to slow the reaction. The fuel is then regarded as spent and must be replaced. The spent rods contain high concentrations of highly radioactive fission products of ^{238}U and ^{235}U. From a health and environmental stand-point, the question of what to do with these and the even more abundant radioactive wastes from military sources has proved to be by far the most difficult problem of the nuclear age. The original goal was to reprocess the waste, remove its uranium and plutonium for reuse, and then safely store the many remaining by-products of radiactive fission for an indefinite period. In the case of the power plants alone, it was anticipated that for each reactor in full operation, some 10 to 60 shipments per year would be made to reprocessing plants; until 1972, specially designed casks were used in shipping, and most shipments were made by truck or rail.

Reprocessing of fuel rods involves dissolving them in concentrated acid and recovering the uranium and plutonium. This procedure leaves a waste solution that remains highly radioactive for as long as a million years. The liquid, which is said to boil "like" a teakettle " was to be temporarily stored in huge tanks until it cools. Reprocessing regularly results in the emission of radioactive gases (for example krypton-85) more long-lived than those emitted during the operation of a power plant. A slight increase in cancer deaths—well under 1 per cent—was expected to result in those exposed to this pollution source.

Regulations provided that the liquid be convered to a solid form within 5 years and shipped to an approved depository within 10 years. What some regard as the main problem with nuclear power occurs at this point: no generally satisfactory way has been found of permanently storing these wastes for as long as thousands of years, let alone hundreds of thousands. Storage of the liquid in tanks has not proved practicable. Not only is it impossible to design a tank that will last for even 100 years, but between 1969 and 1980, 16 cases of leakage from tanks occurred in the United States, permitting the escape of over 1.3 million l (350,000 gallons) of radioactive liquid into the environment.

The impasse over the storage problem and problems associated with reprocessing nuclear fuel in the United States has halted all reprocessing since 1972. One approach to storage is to incorporate the liquid into solid blocks of glass for storage in abandoned salt mines equipped with remote-control television camers for continual monitoring. Not only does salt conduct heat well—the glass-blocks will remain hot for some time—but most salt mines are dry and sustain little damage from earthquakes. Among other proposals is one to drop the blocks of glass into the ocean at some of the deeper and quieter locations or sink them into deep holes bored in the ocean floor. Partly because of the lack of agreement on how and where to store high-level radioactive wastes, by 1985 over 400,000 tons of spent fuel rods had accumulated at power plant in the United States.

Nuclear Accidents. In addition to the operationalc ontamination of the environment, another area of concern is the possibility of a major accident at a nuclear facility, with the potential for deadly contamination of air, soil, and water in a wide area surrounding the accident site. The only accident that has thus far resulted in massive contamination of a large area apparently took place in 1957 in the Ural Mountains, in the Soviet Union. According to a 1980 report of the Oak Ridge National Laboratory, in Oak Ridge, Tennessee, an accidental explosion occurred during nuclear waste processing at a remote Soviet weapons plant, dispersing millions of curies* of strontium-90 and other nuclides, contaminating over 1,000 km,^{2}

* The curies, a unit of radioactivity, is defined as 3.7×10^{10} nuclear disintegrations per second.

including 14 lakes. Like most details of the accident, the number of casualties has been kept secret. Thirty towns were permanently evacuated and subsequently eliminated from Soviet maps, and 60,000 survivors were relocated.

THE HUMAN POPULATION EXPLOSION

If the world population were to stop increasing at this moment or were even to decrease to half its present size, the problems of pollution and ecosystem destruction would still be with us. Thus, solving the serious problem of the world population explosion could not in itself solve the problem of environmental deterioration. However, the rapid expansion of populations in countries with widespread poverty makes it extremely difficult to arrest some types of environmental deterioration in those countries.

The world's human population is indeed exploding. Although Homo sapiens has lived on earth for at least 300,000 years, it was not until 1800 that we attained a population of 1 billion. Then, within only 130 years, the population increased to 2 billion. In only 30 more years 1930 to 1960), a third billion was added, and in another 15 years (1975), fourth! By 1985, the rate of increase had begun to slow to about 1.8 per cent per year.* Nevertheless, the world's human population is expected to each 6.35 billion by the year 2000, a further increase of over 50 per cent. The human population explosion exemplifies the exponential growth capability possessed by all species. A companion principle dictating that when population size exceeds carrying capacity massive mortality results is also being demonstrated by the explosion, but at regional rather than global levels.

The rapid population expansion since 1800 has not been due to an increase in birth rate, that is, to an increase in

* The global decline in the rate of increase means that world population is increasing at a slower rate, not that the world population itself is decreasing.

the number of births per 1,000 population. Rather, it has been due mostly to a sudden decrease in death rate, due to such factors as the development of relatively inexpensive medicines, methods of sanitation, and other means of disease control. Also contributing to the problem is the persistent cultural practice of raising large families, a custom influenced in the past by very high infant and childhood morality rates. Now that in most countries children usually live to have children of their own, populations have soared. Cultural adjustment to sudden change is typically slow, and adjustment to longer life expectancy has been no exception.

DIMENSIONS OF THE PROBLEM

Although the worldwide effect of the decreased death rate has been a population explosion, population growth in numbers is very uneven from one country to the next, with 90 per cent of it occurring in the "less developed" countries. For example, while populations in Japan and the United States are growing at an annual rate of 0.1 per cent, until recently Mexico's has been increasing at a rate of nearly 3 per cent per year, as have the populations of Iran and Iraq. Many of the world's cities, to which people move in attempts to improve their lot in life, are turning into nightmarish jungles with growing shantytowns housing millions upon millions of inadequately fed and clothed people who lack medical care. There is a higher incidence of malnutrition in such cities than in the rural areas from which the migrants came.

FERTILITY CONTROL

As a general rule, the countries in which fertility control has been most effective are those in which the central government has backed a programme of public information about and encouragement in family planning, usually by contraception. For example, the government of Mexico has claimed that, through such programmes, it reduced the country's rate of growth from 3.4 per cent to 2.7 per cent per year between 1977 and the end of 1980. The problem of limiting population growth is complex and includes cultural, religious, political, and economic factors, any one of which can block solutions. Moreover, birth control programmes alone, without education, health care, and economic opportunity for the majority of a country's population, have rarely been effective. Since the beginning of the 1970s, however, there has been a remarkably sharp drop in fertility in some areas of the world, largely in response to the changes in public attitudes that began in the early 1960s. This decrease has coincided with the rapidly growing use of oral contraceptives and an increase in induced abortions.

THE GLOBAL FOOD PROBLEM

Despite the declining fertility of the remaining arable topsoil, world food output is expected to keep pace with population growth into the next century. Stating the matter this simply, however, is deceptive. In many cases, the increased food production, while it has prevented massive starvation, has depended heavily on expensive fertilizers, pesticides, and irrigation. It is feared that this may only set the stage for even greater population crashes in the future. And although overall food production has kept pace with overall population growth, little food usually reaches those who need it the most. The famines that the world experiences are not general, but regional. Millions have starved: in Biafra in the 1960s, in Bangladesh in 1971, in the sub-Sahara Sahel in the mid-1970s, in Cambodia in 1980, and in Uganda, Ethiopia, and Somalia in the 1980s. While surplus food can almost always be obtained by those who can afford to buy it, little of it reaches the starving poor who cannot.

Thus, the world's ability to raise enough food to feed everyone could not feed the over 16 million refugees from war and famine in the early 1980s and cannot feed the 500 million people who are hungry every day of their lives. Although global relief programmes may often compensate somewhat for these situations on an emergency basis, populations continue to grow and poor people are increasingly cultivating the remaining marginal lands. Both trends forbode unprecedented ecological disasters for the future. There appears to be little question that world population will exceed 6 billion by the year 2000, with the increase occurring largely in those areas already suffering most from hunger, malnutrition, and starvation, namely, parts of Asia, Latin America, and Africa.

THE FUTURE

Is there hope for the future? It is easy to feel overwhelmed by the magnitude of the many environmental problems that confront us—to feel that one person is helpless against all the forces that threaten our planet. Although one person might indeed be helpless, literally hundreds of thousands of people are attempting to solve these problems. Moreover, they are achieving success on many fronts; to recount their achievements would fill at least another chapter. The environmental crisis has many facets, and many specialized groups have arisen to meet them. Some organizations are international in scope; some are local. Some appeal to the public, some lobby legislatures, and some act directly by buying up tracts of land to preserve them from reckles development. Some have broad politico-economic perspectives in which respect for the environment is an integral part.

In a number of cases, the action of citizens' groups, of industries, or of governments at the urging of citizens has halted needless destruction of ecosystems of restored severely damaged areas. Although such cases are exceptional and there is much work yet to be done, the success that has been achieved is proof that it is possible. Alternatively, to yield to despair is to guarantee disaster. For those interested in a never-ending battle, we suggest joining one or more of the many agencies and citizen groups that seek to divert us from the dangerous path down which our civilization is heading.

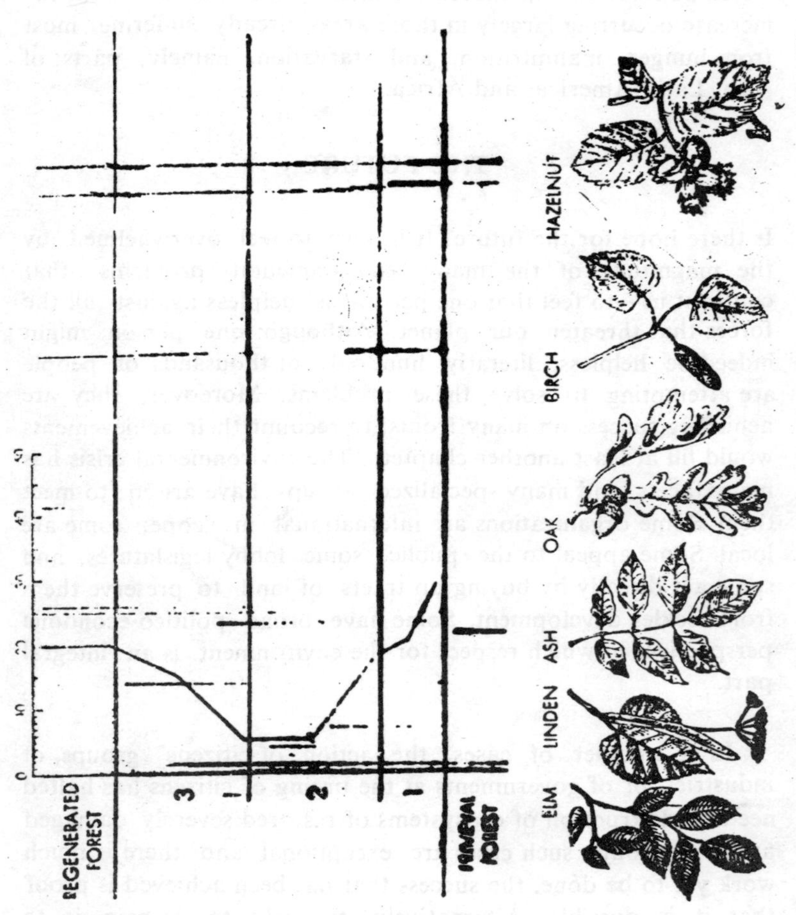
HAZELNUT
BIRCH
OAK
ASH
LINDEN
ELM
REGENERATED FOREST
3
2
1
PRIMEVAL FOREST
0
10
20
30
40
50

17

Flies that Parasitise Vertebrates

Diptera are the only insects that regularly infest the bodies of vertebrates; such an infestation is termed myiasis. The maggots feed either upon ingested food or upon healthy or necrotic tissue of the living host. In nearly all cases development proceeds, at least for a while, inside the host. By no means all examples of myiasis are examples of parasitism. Maggots feeding on dead wound tissue or on faecal matter in the rectum are certainly not behaving as parasites. To omit mention of these non-parasitic forms, however, would render more difficult an understanding of the evolution of myiasis.

Dipteran larvae are not infrequently accidently eaten by man, and as the higher Diptera have a cuticle which is very resistant to penetration by toxins, they are quite often carried passively right through the alimentary tract to emerge, still living, from the anus. The cheese-skipper, the larva of Piophila casei, which is often a serious pest in cheese and meat, has frequently been reported to pass live through the human gut. Infestations of this type are termed pseudomyiasis since the larvae concerned do not develop inside the 'host', but in practice it is often difficult to distinguish between pseudomyiasis and cases of facultative intestinal myiasis.

A curious case involving Megaselia scalaris, a species of Phoridae whose larvae feed in a wide variety of decomposing organic materials (including emulsion paint!), is quoted by Patton and Evans (1929). A man living in Burma passed in his faeces, over a period of about a year, Megaselia larvae of various ages, puparia, and even adult flies. It is believed that larvae had been accidentally ingested with food, but the extended duration of the infestation, persisting long after the patient was placed under medical supervision, together with the occurrence of adult flies, has led to the suggestion that the flies bred and passed through a number of generations within the gut. Zumpt (1965) dismisses this suggestion, perhaps rightly, as an impossibility, although it is known that some Phoridae pass several generations on interred corpses.

Myiasis can be either facultative or obligatory. Flies whose larvae are free-living in decomposing organic materials may occasionally gain access to the body of a vertebrate and there develop, at least for a while, on faeces, suppurating wounds, or the like. Several species of Calliphoridae and Muscidae exhibit this facultative myiasis. Obligatory myiasis producing flies live as larvae only upon vertebrates, either externally like Auchmeromyia and Neottiophilum, or internally like warble flies and bot-flies (Oestridae, Gasterophilidae). Flies such as Hippoboscidae, which feed upon vertebrates only when adult, do not come into the category of myiasis producers.

The subject of myiasis is comprehensively reviewed in a monograph by Dr. F. Zumpt (1965) of the South African Institute for Medical Research. In this work only species from the Old World are considered, and 63 are listed as causing facultative myiasis or pseudomyiasis, and 106 as causing obligatory myiasis. These species are distributed among the following families and genera:

Facultative Myiasis

Anisopidae: Anisopus

Psychodidae: Psychoda

Phoridae: Megaselia

Sytphidae: Eristalis

Piophilidae: Piophilus

Ephydridae: Teichomyza

Muscidae: Musca, Muscina, Stomoxys, Ophyra, Fannia

Calliphoridae and Sarcophagidae: Lucilia, Calliphora, Protophormia Chrysomya, Sarcophaga, Wohlfahrtia

Obligatory Myiasis

Neottiophilidae: Neottiophilum

Chloropidae: Batrachomyia

Muscidae: Passeromyia

Calliphoridae and Sarcophagidae: Lucilia, Pachychoeromyia Auchmeromyia, Cordylobia, Booponus, Elephantoloemus, Protocalliphora, Chrysomya, Wohlfahrtia. The New World genus Callitroga also belongs to this group.

Gasterophilidae: Gasterophilus, Gyrostigma, Cobboldia, Platycobboldia, Rodhainomyia Ruttenia, Neocuterebra. Plus Cuterebridae: Cuterebra, Dermarobia of the New World.

Oestridae: Pharyngomyia, Cephenemyia, Pharyngobolus, Tracheomyia, Kirkioestrus, Rhinoestrus, Oestrus, Gedoelstia, Cephalopina, Portschinskia, Oestromyia, Strobiloestrus, Pavlovskiata, Pallasiomyia, Oedemagena, Przhevalskiana, Hypoderma.

Zumpt traces the evolution of myiasis, as indicated by the habits of flies living today, and recognises two roots, the saprophagous root and the sanguinivorous root. The saprophagous root originates with a larva of rather generalised feeding habits able to live on a diversity of decaying organic material and sometimes including in its diet the dead tissue associated with septic wounds on vertebrates.

Such an insect is the larva of the greenbottle, Lucilia sericata, which most often feeds on carcasses, but also occasionally on purulent wounds. In wounds it consumes only dead tissue, and may even have a beneficial effect in cleansing the affected region. In fact L. sericata maggots had a limited medical use up to about 1930. Larvae reared under sterile conditions were used to speed the healing of septic wounds (Brumpt 1933). In additian to the removal of dead tissue, Stewart 1934) claims that calcium carbonate is secreted by the larva and that this stimulates phagocytosis by creating an alkaline environment. Simmons (1935) found that larval secretions of L. sericata were themselves bactericidal, and this was later confirmed in other species. The larva of the bluebottle Protophormia terraenovae produces a secretion which inhibits the growth of bacteria, an obvious advantage to a larva that lives in an environment which could be destroyed by bacteria (Pavillard and Wright 1957).

Lucilia cuprina may also feed at suppurating lesions. It is closely allied to L. sericata, but unlike that species its diet is not retricted to dead tissue, and it will ingest healthy tissue surrounding a wound, thereby extending the damaged area. The final step along this evolutionary path has been taken by flies whose larvae are incapable of developing in carcasses and other types of decomposing organic matter, and rely entirely upon sores on vertebrates to supply them with their nutritional requirements. Chrysomya bezziana, Callitroga hominivorax (=americana), and the larviparous Wohlfahrtia magnifica are three such obligate parasites C. bezziana and W. magnifica both belong to genera whose other species are nearly all feeders, as larvae, in faeces, carcasses, or facultatively in wounds. C. bezziana will oviposit in very small sores such as midge or tick bites, and here the first stage larvae feed upon blood and lymph, Later, the second stage larvae invade healthy tissue and the final, third stage larvae are almost completely embadded in living tissue before they are fully fed and drop to the grourd to pupate. The maggots or screw-worms of C. bezziana attack man and domestic animals in Africa and the Oriental region, not infrequently turning minor injuries into mortal wounds.

Callitroga (=Cochliomyia) is a screw-worm of the New World, and it achieved fame as a result of being decimated by a control procedure termed the sterile-male technique. The species was reared in vast numbers in the laboratory and late pupae were irradiated by X-rays. This had the effect of sterilising the males, or at least of inducing lethal mutations in the sex cells, but at the same time it did not alter the males' vitality or longevity. In 1954 and 1955 irradiated males were released weekly on the island of Curacao in such quantities that they outnumbered normal males. They competed for females with the native males but, of course, were incapable of fathering offspring. The female C. hominivorax will copulate once only, whether fertilised or not by the union, and so the large numbers of sterile males had a very detrimental effect upon the population. Males mate several times. By this technique, Curacao was cleared of screw-worms within five months (Bushland 1959). Knipling (1955) showed on theoretical grounds that a population of 20 million insects, 10 million of which are healthy males, can be virtually wiped out in four generations by releasing 20 million sterile males in each generation. The sterile male technique was used to rid Florida of its screw-worms. Success was achieved in two years at a cost of 8 million dollars which compared favourably with the annual to million dollars' worth of damage inflicted by screw-worms (Weidhaas, Schmidt, and Chamberlain 1962). In England the techniques has been employed against the green-bottle, Lucilia sericata, on the island of Lindisfarne but in this case control was not achieved (Macleod and Donnelly 1961).

A more specialised obligatory parasite is Lucilia bufonivora, a species which attacks toads and other amphibians. Eggs are laid on the skin and, on hatching, the young larvae migrate to the nostrils or, occasionally, the orbits. Here they cause extensive lesions and the host is generally killed within a week, the larvae sometimes completing their feeding upon the corpse.

The thin, moist, amphibian skin allows L. bufonivora to gain a foothold on previously uninjured hosts, the toad's mucus providing a substrate for the fly larvae to move in without the risk of desiccation. In contrast, blowflies attacking mammals require at least a small pre-existing sore or wound in which an infestation can be initiated. Much research has been done on sheep strike, that is wound myiasis in sheep, and it is known that a succession of species may be involved (Haddow and Thomson 1937, MacLeod 1937). Sheep strikes are generally found, if not in actual injuries, on those regions of the host where the fleece is frequently moistened with urine or faeces. Larvae of primary strike-flies fuçh as Lucilia sericata and Protophormia terraenovae in Britain, and Lucilia cuprina and Calliphora stygia in Australia, will live in soiled fleece, where bacterial activity causes inflammatian of the host's skin accompanied by lymphal exudations. The larvae feed at the surface of the epidermis which is irritated, causing the stricken sheep to rub or scratch the affected part. This creates larger lesions which can be exploited by more maggots, either of the primary species or of secondary strike-flies. The odour of putrefaction from the primary strike attracts secondary strike-flies like Lucilia caesar, L. illustris, Calliphora erythrocephala, C. vomitoria and several others. Finally, tertiary strike-flies such as species of Musca and Fannia may afflict severely stricken sheep. Large numbers of sheep are destroyed by these flies and the commercial value of the wool of many more is much reduced. The problem of sheep strike is particularly prevalent in Australia, the dense fleece and wrinkled skin of Merino sheep providing a humid environment and being very liable to soiling.

Zumpt (1965) adds a small branch to his saprophagous root of myiasis. Many musciform flies oviposit ar larviposit in vertebrate faeces, occasionally even as faeces are leaving the anus. Such habits could lead to rectal myiasis, in which eggs or larvae are laid around the anus and the larvae crawl into the rectum to complete their feeding inside the body. Musca domestica and Sarcophaga haemorrhoidalis have been recorded as causing rectal myiasis. Similarly, larvae of species of

Psychoda, Musca, Fannia, and Calliphora may enter the urethra and sometimes even penetrate as far as the bladder. No flies are known to be obligatory causers of rectal or urinogenital myiasis.

Zumpt's sanguinivorous root of myiasis, like the saprophagous root, begins with a relatively unspecialised feeder in decomposing organic matter. Some species, Muscina stabulans for example, feed saprophagously during their early larval stages, but later develop a predatory tendency, attacking and eating other maggots sharing their habitat. Zumpt suggests that such a larva, with mouthparts capable of piercing other larvae, could, if it lived in the nest of a bird or mammal, pierce the skin of a vertebrate occupant of the nest and take a blood meal. This facultative ectoparasitism could subsequently develop into on obligatory relationship such as is now the case with species of Neottiophilum, Passeromyia, and Protocalliphora in birds' nests, and species of Pachychoeromyia and Auchmeromyia in mammalian dwelling places.

Ectoparasitism by larvae of endopterygotes is most exceptional, and in these flies the habit could have developed only in circumstances that allowed the parasite close and prolonged proximity with a host in an almost immobile condition.

The genus Neottiophilum contains a single species, the European N. praestum, whose larva sucks the blood of nestling passerine birds. It has a single generation a year, and this is synchronised with its hosts, adults hatching in spring to coincide with the birds' breeding season (Keilin 1924). Pupation occurs in the nest. Passeromyta heterochaeta lives as a larva in nests of swallows, starlings, weaver birds, and others, mainly in Africa, taking blood meals from the nestlings. For the first two larval instars it lies on the host's body, attached to the plumage by posterior filaments. An allied species, the Australian P. longicornis, has developed a more intimate association with its hosts in that after initially feeding externally like P. heterochaeta, it later burrows into the nestling's skin and forms a swelling with only the posterior spiracles pratruding for respiratory purposes. A similar evolutionary trend is suggested by the hapits of the European species of Protocalliphora. P. azurea, like Passeromyia heterochaeta, is an ectoparasitic blood-sucker of nestling birds, particularly tits and warblers. So also are P. chrysorrhoea in sand martins' nests, P. peusi in crows' nests, and P. falcozi which has been found most often in great tits' nests. However P. lindneri, which attacks ground-nesting passerines, and P. braueri whose principal host appears to be the house sparrow, parallel P. longicornis in burrowing subcutaneously and forming boils. They fully grown larvae of all of these species pupate in the nest material. Heavy infestations may kill the host.

Other Calliphoridae have become obligate parasites of mammals. Auchmeromyia and Pachycheromyia feed on the blood of African mammals and are ectoparasites behaving like Passeromyia heterochaeta. The single species of Pachychoeromyia, P. praegrandis, and the five described species of Auchmeromyia, are all associated with warthogs and aardvarks, sometimes also with the aardwolf. Warthogs and aardwolves frequently make use of aardvark burrows as dwelling places, but the animals are not closely related taxonomically. A. luteola, the Congo floor-maggot, in addition regularly includes man's blood in its larval diet. The female fly lays up to 300 eggs singly in dry, sandy soil in shaded places, sometimes on the floors of native huts. All larval stages are blood-suckers, and as parasites of man they depend upon people sleeping on the bare ground or on mats. Feeding takes place at night, usually every night, when the feeding maggot scrapes at the host's skin with its mouthparts until the blood capillaries are lacerated.

Parasites that form boils on mammalian hosts, equivalent to the bird parasite Passeromyia longicornis, are found in the genera Cordylobia, Booponus, and Elephantoloemus. Cordylobia anthropophaga, the tumbu fly, is undoubtedly the best known species (Blacklock and Thompson 1923). Eggs are laid in sandy soil, usually in the vicinity of urine or faeces, and at first the larvae live just below the soil surface. When a man walks close to where they are lying, they come on to the surface and wave the anterior parts of their bodies about. Should they contact skin, usually of the feet, penetration is rapidly effected, and within a very short time only the posterior spiracles of the maggot are left exposed to the outside world. A boil or abscess forms about the maggot and it rapidly grows. After perhaps as little as eight days the larva may be fully grown. It now leaves the boil and pupates in the ground. A well as man, the tumbu fly will attack rats, dogs, monkeys, and several other mammals, even birds, but survival in non-human hosts is uncertain. C. rodhaini has also been recorded as attacking man but the rate of survival is low, and its most usual hosts are rodents. The only other described species of Cordylobia, C. ruandae, may be a specific parasite of the mouse Grammomys dolichurus. All three Cordylobia species are known only from Africa. Booponus is an oriental genus of four described species which are parasites of Artiodactyla. Like Cordylobia, the larvae live in skin boils, but an interesting point of difference between the two genera is that Booponus species oviposit on the hairs of their hosts, not on the ground. This is necessary because Artiodactyla do not frequent habitual resting places. Elephantoloemus indicus, the final species in this group of mammalian parasites, is a specific parasite of the Indian elephant.

The Chloropidae is a family with diverse larval habits. Most are plant-feeders, often gall-formers on grasses, and included amongst them are agricultural pests such as the frit fly and gout fly. Others feed upon rotting vegetable material, and some are predators of aphids, beetles, and spiders' eggs. However Batrachomyia, an Australian genus, is unique in its feeding habits. There are ten described species, all of which are probably obligate parasites of frogs and toads, possibly host-specific, living as final-stage larvae in subcutaneous swellings. Infested hosts often die.

More specialised than the preceding families as larval parasites of vertebrates are the Gasterophilidae and Oestridae, bots and warbles, two allied families all of whose species are obligate parasites.

Gasterophilidae

The adults of this family are large and attractive flies whose larvae, as their family name implies, are 'stomach lovers' or stomach bots. The mouthparts of the adults of both gasterophilids and oestrids are rudimentary.

Gasterophilus is the principal genus. There are nine species, and they chiefly inhabit the Ethiopian and palaearctic regions. They are all parasites of horses, asses, and zebras: occasional parasitism of man does not persist beyond the first larval instar. Although the species tend to share the same hosts, details of their life histories differ. The female G. pecorum lays her eggs, which may number over two thousand, in rows upon grasses and other plants. The larvae do not batch until the eggs are eaten by the host; they may remain viable on vegetation for several months. In contrast, G. haemorrhoidalis, G. intestinalis and G. nigricornis lay their eggs upon the hairs of their hosts. G. haemerrhoidalis oviposits upon the lips, where moisture stimulates their hatching, and the young larvae burrow into the epidermis of the lip and into the mouth. G. intestinalis, on the other hand, lays it eggs upon the legs or back of its host, hatching being stimulated by friction when licked, and the young larvae penetrate the tongue. G. nigricornis selects the cheeks for egg-laying, and it differs from the other two species that its eggs apparently hatch spontaneously without stimulation from the host. The young larvae enter the mouth by burrowing into the epidermis at the corner. The sites occupied by the later larval stages in the alimentary canal of the host also differ specifically. G. pecorum occupies the soft palates and adjacent areas in its second larval instar, and the stomach in its third (final) instar. The second instar larva of G. haemorrhoidalis moves to the stomach and duodenum, where it moults, and the third instar larva migrates to the rectum of its host. The second stage larvae of both G. intestinalis and G. nigricornis pass from the mouth (tongue and check respectively) to the stomach or duodenum, where they become third stage larvae. In all species, fully grown larvae pass out of the anus and pupate on the ground in the surface layers of soil.

Whereas Gasterophilus confines its attentions to the Equidae, Platocobboldia loxodontis, Rodhainomyia roverei, and Cobboldia elephantis live as final stage larvae in the stomachs of elephants, the first two being specific parasites of the African elephant whilst the last-mentioned is known only from the Indian elephant. Only the life cycle of Platycobboldia has been elucidated. The eggs are laid about the bases of the tusks, and the first stage larvae enter the mouth. The fully grown larvae move about freely inside the stomach, unlike those of Gasterophilus which are attached to the gastric epithelium. Another interesting deviation from Gasterophilus is that fully grown Platycobboldia larvae do not pass out of the host's anus, but instead migrate up the oesophagus to congregate beneath the tongue prior to being evacuated from the mouth. Pupation again occurs in the soil.

Another distinct group of the Gasterophilidae, comprising the three known species of Gyrostigma, is parasitic on rhinoceroses. As far as is known, their eggs are laid upon the host's head and the larvae are attached to the stomach wall, passing out of the anus when fully grown. Gyrostigma pavesii and G. conjungens parasitise the African rhinoceroses, Ceratotherium simum and Diceros bicornis respectively, whilst G. sumatrensis is known only from larvae collected from a Sumatran ryinoceros (Didermoceros sumatrensis). All of these flies are large, and several of them are beautifully coloured, Platycobboldia loxodontis has smoky wings, an orange-colhured head, and metallic-blue thorax and abdomen. Some species of Gasterophilus are very hairy and resemble bumble bees.

Two other species placed by Zumpt (1965) in the Gasterophilidae are biologically atypical of the family. Both are parasites of the African elephant but their larvae do not dwell in the alimentary tract of their host. Neocuterebra squamosa larvae are found in the feet and Ruttenia loxodontis larvae develop in boils in the skin of various regions of the body. As Zumpt points out, the latter species occupies in the African elephant the ecological niche filled by the calliphorid Elephantoloemus indicus in the Indian elephant.

Cuterebridae

These are New World botflies that cause dermal myiasis of man, rodents, and lagomorphs. There is a fairly clearly defined host-specificity in the family which is probably of ancient origin parasitic upon rodents. Cuterebra latiforns, a parasite of the woodrat (Neotoma fuscipes), has been investigated in California by Catts (1964), Males are attracted to the summits of hills, where they exhibit territorial behaviour and become spaced out. Females fly to the hill-tops to mate and then disperse. Eggs are laid, in small groups of up to ten, near woodrat nests. One fly is recorded (Radovsky and Catts 1960) to lay 270 eggs. Egg hatching is sometimes stimulated by mechanical disturbance. The young larvae adhere to the substrate by a caudal 'holdfast' and they respond to the presence of a mammal by making swaping, questing movements of their onterior regions. Entry into a host is normally effected through a body orifice. In the laboratory, woodrats have been infected *via* their nares after smelling sticks on which were C. latifrons larvae. Other rodents were also readily infected in this manner, but development of the fly larvae did not usually proceed to completion. Laboratory mice died at an early stage of infection.

Cuterebra larvae live in subdermal cysts which communicate with the exterior by a pore. The fully grown larves its cyst to pupate in the ground. A frequent site of infestation by Cuterebra is the inguinal region; male hosts of C. emasculator may have their testes destroyed by the parasite. C. angustifrons is chiefly a parasite of the white-footed mouse (Peromyscus leucopus) and it has a deleterious effect upon blood composition and movement of its host (Dunaway *et al.* 1967).

Dermatobia hominis the human botfly of Central and tropical South America, does not itself search for the where-bouts of a host. It is phoretic and glues its eggs, of which it produces many, to the abdomens of blood-sucking flies especially mosquitoes. In this way a batch of up to one hundred eggs is transported to the host, which is often a cow but may be a man. How Dermatobia evolved the habit of phoresy, so different from Cuterebra, is unknown. The larvae emerge is response to warmth and each usually enters the host through a hair follicle. The larva causes a boil or cyst to form about it in the subcutaneous tissue, and here it attains full size in about six weeks, remaining all the while in contact with the atmosphere by a small aperture in the centre of the boil. The fully grown larva eventually emerges through this aperture to pupate in the soll. The larva of D. hominis behaves in many ways like a larva of a warble fly (Hypoderma) but unlike Hypoderma emerges not migrate from its original site of entry.

Oestridae

The Oestridae are sometimes classified as a tribe of the Tachinidae (*e.g.*, van Emden 1954), Zumpt (1965) grants them family status and recagnises two subfamilies, Oestrinae and Hypoderminae, which have quite distinct larval feeding sites. The larvae of Oestrinae feed in the naso-pharyngeal regions of ungulates, both Perissodactyla, and Artiodactyla, with single species attacking marsupials and elephants. The Hypoderminae parasitise Artiodactyla, rodents, and lagomorphs, and as their name suggests, they live subcutaneously, stimulating the formation of small boils.

Among the Oestrinae, the best known species is Oestrus ovis the sheep bot, a cosmopolitan species now but probably of palaearctic origin. Like others in the subfamily it is larviparous, depositing first-stage larvae in the nostrils of the host. The maggots develop in the nasal and frontal sinuses feeding on mucus and blood. When fully grown, they are sneezed out by the host and pupate on the ground. Sheep and goats are the principal hosts of O. ovis but occasional attacks on man have been recorded. Curiously, when attacking man, the larvae are laid; not in the nostrils, but in the eyes; their development does not usually progress beyond the first instar. Other Old World

Oestrinae have similar life cycles to O. ovis. Their hosts, according to Zumpt (1965), are as follows:

Deer; Cephenemyia (4 species), Pharygomyia (1)

Antelopes, Gazelles: Pharyngamyia (1), Oestrus (4), Kirkioestrus (2), Rhin oestrus (2), Gedoelstia (2)

Giraffe: Rhinoestrus (1)

Camel: Cephalopina (1)

Sheep, Goats, Ibex, Tur: Oestrus (2), Rhinoestrus (1)

Hippopotamus, Bushpig, Warthog: Rhinoestrus (3)

Horses, Zebra: Rhinoestrus (4)

African Elephant: Pharyngobolus (1)

Kangaroo: Tracheomyia (1)

The Oestrinae probably originated in the palaearctic region but are now cosmopolitan. Cephenemyia trompe is a circumpolar species parasitising reindeer in the palaearctic and caribou in the nearctic.

The best know representatives of the subfamily Hypoderminae are Hypoderma species large flies often resembling bumble bees, that pass their larval stages as parasites of Bovidae and Cervidae. The common warble fly of cattle is H. bovis. The female fly lays its eggs on the hairs of a cow, particularly on the belly and thighs. The act of oviposition is performed very rapidly. On hatching, the young larvae crawl down the hairs to the skin, which they penetrate, and then embark upon a prolonged migration inside the host's bady which lasts for about four months. At first the larvae follow the course of nerves until the spinal cord is reached. Here they remain a while before burrowing through fat and muscle towards the skin of the back. A small breathing hole is opened to the outside and the host's tissues swell about each warble larva. In these dorsal swellings the larvae grow and moult twice in three months before becoming fully grown. They larve the swellings to pupate on the ground. There is one generation during a year. The larva of Hypoderma, like that of Protophormia terraenovae, produces a secretion which has bacteriostatic properties (Beesley 1968).

The life cycle of H. lineatum differs from that of H. bovis in that the first instar larva migrates through the oesphageal submucosa instead of the spinal nerves. This species is another parasite of cattle, and it lays its eggs mostly on the heels. A few cases of human infestation have been recorded, and although in such cases the iarvae are unable to complete their development, they may destroy the eyes.

H. bovis and A. lineatum adults both cause cattle to 'gad'. It is as if cattle are aware of the suffering they will endure long after the flies have laid their eggs.

Other species of Hypoderma are parasites of deer, and in most cases they are host-specific. Species of the palaearctic genera Pavlovskiata, Pallasiomyia, and Przhevalskiana attack Asian antelopes or gazelles, sheep, and goats, whilst African antelopes are parasitised by species of Strobiloestrus.

Oedemagena tarandi is the reindeer warble fly, a host-specific species that as a larva migrates to the back of its host through the muscles of the hindquarters (Hadwen 1926). The herd of reindeer introduced to the Cairngorms in Scotland were screened for warbles before being released. However, soon after liberation they were found to be infested. The parasite was not O. tarandi but Hypoderma diana, normally a red-deer parasite. The H. diana larvae in reindeer behaved abnormally, making several punctures in the skin, aud some reindeer died as a result of infestation (Kettle and Utsi 1955).

Distinct from the group of genera whose hosts are ungulates is another section of the subfamily parasitic upon rodents. These species also cause subcutaneous boils, either on the back or belly, and because they are no smaller than the ungulate parasites, a host can usually support only a single larva at a time. Palaearctic species of Oestromyia, Oestroderma, and

Portschinskia attack marmots, pikas, voles, and mice. Their biology is poorly known.

Canthariəsis and Scholechiasis

Files are not the only insects whose larvae may enter the bodies of vertebrates. Lepidopterous caterpillars on vegetables may be eaten accidently by man, but it is rare for them to survive long in the vertebrate gut. Not so casual is the occurrence of larvae of Tinea vastella in the horns of African ungulates. This species is probably an obligatory parasite. Such lepidopterous infestation of vertebrates is termed scholechiasis.

Beetles figure more frequently than Lepidoptera among recorded cases of insect infestations of man. Adult dung beetles (Scarabaeidae) sometimes enter the human recturn by way of the anus. These beetles lay their eggs on vertebrnte faeces, and it is perhaps not surprising that some follow to the end their search for oviposition sites. Species of Onthophagus in India and Ceylon seem especially prone to this habit, and children are the principal victims. Many Dermestidae feed on dried animal skins, and occasionally larvae have been found making feeding galleries in the skin of live, nestling pigeons (Paulian 1943). Coleopterous infestation of vertebrates is termed canthariasis.

18

Signs and Symptoms of Parasitic Disease

Very few of the signs or symptoms evoked by infection with parasitic organisms can be said to deserve that much-abused term pathognomonic. There are but a limited number of ways in which the body is able to react to altered conditions, and therefore it is not surprising that few of these reactions are specific. However, the presence of certain signs and symptoms should alert the clinician to corresponding diagnostic possibilities, while various constellations of symptoms, or syndromes, are to a greater or lesser degree diagnostic.

It must be emphasized that the following is not intended as a complete differential diagnosis of any of the symptoms discussed. Limitations of space do not permit even mention of the various non-parasitic etiologies of many of these conditions.

Abdominal pain. Crampy abdominal pain may characterize amebic colities, with tenesmus if the ulcerations involve the rectal area. Diarrhea or dysentery is usually present when the infection is symptomatic. If hepatic abscess develops, pain is usually felt in the right upper quadrant (left, if the abscess involves the left lobe), and may be referred to the scapular area.

Pain in giardiasis is usually mild but may occasionally be severe; it is usually crampy and may be accompanied by steatorrhea and a full-blown malabsorption syndrome. Pain is seldom present in intestinal worm infections. Intestinal or biliary obstruction (the former primarily in smaller children) can be the result of Ascaris infection, with signs and symptoms which mimic obstruction from any other cause. Strongyloides stercoralis, invading the mucosal wall, may cause a severe duodenitis or jejunitis, with symptoms suggestive of duodenal ulcer disease. A moderate to heavy eosinophilia generally accompaines either these worm infections.

Abscess, amebic. Amebic invasion of the liver is characterized by tenderness and enlargement of that organ, progressive malaise, an irregularly spiking fever with night sweats, leukocytosis, elevation and fixation of the right diaphragm, and sometimes development of a right lower lobe pneumonitis. With abscess formation, pain becomes more intense and may be referred to the tip of the right (less commonly) the left scapula.

Abscess, filarial. Abscesses may develop spontaneously or appear shortly after antifilarial treatment is begun. They occur along the course of lymphatics or at lymph nodes and may be distinguished from pyogenic abscesses by the fact that they are generally sterile when first opened and that fragments of the adult worms may be found in the abscess drainage.

Anemia. Most frequently associated with malaria, hookworm, and broad fish tapeworm infections, anemia may be seen in kala-azar, trypanosomiasis, sclaistosomiasis, fasciolopsiasis and trichuriasis. In falciparum malaria the red cell count may fall to 2.5 to 4.0 million in cases of averege severity, and under 1.0 million in severe infections. Anemia is usually not severe in vivax malaria and is still less pronounced in quartan infection. The characteristic microcytic hypochromic anemia of hookworm infection is the result of blood loss and is thus proportional to the severity of infection, although adequate dietary intake of iron may prevent its development in light or moderate infections. The small amount of blood ingested per Trichuris makes anemia rare in other than massive infections. Persons infected with Diphyllobothrium latum may develop a macrocytic hyperchromic anemia on the basis of vitamin B_{12} deprivation if the worm is attached to the jejunal wall. In kala-azar, and possibly also in Chagas' disease, anemia may be a consequence of proliferation of infected reticuloendothelial cells in the bone marrow. In the other trypanosomiases, schistosomiasis, and certain intestinal helminthic infections such as fasciolopsiasis, anemia may result in part from nutritional causes.

Appendicitis. Amebic ulceration involving the cecal area or appendix may simulate acute appendicits; surgical intervention when there is extensive cecal ulceration may be disastrous. Ascaris may block the lumen of the appendix and give rise to appendicitis; this is reported also for Trichuris. Angiostrongylus costaricensis infection may also mimic appendicitis.

Ascites. Circumoval tissue proliferation leading to extensive fibrosis of the liver in Schistosoma mansoni and S. japonicum infections may lead eventually to a condition clinically very similar to Laennec's cirrhosis, with portal hypertension, splenomegaly and ascites. Although hepatomegaly and splenomegaly occur early in kala-azar, ascites is uncommon. It may occur in chronic cases, probably secondary to a nutritional cirrhosis.

Asthma, bronchial. Asthmatic attacks may occur is Ascaris infection during the stage of migration through the lungs, or later in the course of the infection because of hypersensitization to the absorbed worm antigens. Asthma is not uncommon in visceral larva migrans infections, in which there is usually an accompaning hepatomegaly and marked eosinophilia. See also tropical eosinophilia, under Eosinophilia.

Blackwater Fever See *Hemoglobinuria*

Calabar swellings. Circumscribed subcutaneous swellings are seen in loiasis. They are usually intensely pruritic and may be quite painful if they develop in areas where there is little loose subcutaneous tissue. They appear rapidly, developing within an hour or so to a diameter of several centimeters, and persist of several days, or if in an area subject to repeated trauma may last for a week or longer.

Calcifications, cerebral. Calcification of areas of intracerebral infection in congenital toxoplasmosis may be seen no X-ray, and when intracerebral calcifications are round in a patient who has chorioretinitis, a diagnosis of toxoplasmosis is highly probable. Calcified cysts of Cysticercus cellulosae within the brain may be distinguished by their size and uniform rounded or oval shape.

Chagoma. The hard, reddened, raised primary lesion in Trypanosoma cruzi infection usually develops on the head or neck and sometimes on the abdomen or limbs, and may persist for two or three months. Although it may occur on or about the eye, the chagoma is not to be confused with the unilateral palpebral edema of Romana's sign (q.v.).

Chorioretinitis. Infection of the retina and choroid by Toxoplasma or (rarely) Entamoeba histolytica may produce visual disturbances, which can be profound if the macula is involved. On ophthalmoscopic examination, a grayish or yellow white area surrounded by exudate is seen early; with healing this leaves a white atrophic patch bordered by pigment deposits.

Chyluria. The formation of lymphatic varices, consequent upon repeated attacks of filarial lymphangitis and obstruction of lymphatic drainage, may lead to the passage of lymphatic fluid in the urine, if varices rupture into any part of the urinary tract. Chyluria usually occurs in attacks lasting a few days; the urine way have a milky white colour and contain microfilariae.

Coma. The sudden onset of coma in a patient known to be suffering from falciparum malaria, or in an apparently healthy person in, or recently returned from, a malarious area, should always suggest cerebral malaria and requires emergency treatment. While most common as a complication of falciparum malaria, coma may occur with other types of malaria. In African trypanosomiasis, coma develops after a protracted period of increasi.ıgly severe symptoms of meningoencephalitis. It is also seen in primary amebic meningoencephalitis, in which a history of rapidly developing fever, meningeal signs, confusion and coma, and of recent swimming or diving in fresh water will often be elicited.

Cerebral cysticercosis may be a diagnostic consideration in persons of Mexican or Latin American origin, or of long residence in those areas, who present with a variety of neurological symptoms including headache, alteration of consciousness, focal seizures, coma and internal hydrocephalus. Diagnosis may be difficlut, and eosinophilia (either peripheral or of the spinal fluid) is not always present. Serological tests are helpful.

Conjunctivitis. Chronic conjunctivitis is seen in onchocercal infections, with hyperpigmentation of the conjunctiva, photophobia and gradual development of corneal opacities; acute exacerbations of conjunctivitis and photophobia may be associated with attacks of onchocercal dermatitis (q.v.). The sheep botfly, Oestrus ovis, may lay its eggs in the conjunctivae, and development of the larva in the conjunctival sac is accompanied by considerable pain, localized swelling and conjunctivities.

Convulsions. Focal convulsive seizures of the jacksonian type are seen in a number of parasitic infections that involve the central nervous system. These are discussed under Neurologic Symptoms. Convulsions also may be seen in the malarial paroxysm, in acute toxoplasmosis occuring in newborn children, and in Ascaris infection in children.

Dermatitis. Dermal leishmanoid is a secondary cutaneous manifestation of Leishmania donovani infection, occurring a year or so after supposedly successful treatment of kala-azar. The lesions may be flattened or depressed depigmented macules, or erythematous nodules which, on the face, often occur in a butterfly distribution reminiscent of lupus erythematosus. Leishmanial amastigotes are found in the lesions.

Penetration of schistosome cercariae through the skin causes a localized edema and pruritus, mild in the case of the human schistosomes, more severe in the "swimmer's itch" caused by bird schistosome cercariae. A similar localized pruritic reaction occurs with penetration of Strongyloides larvae through the skin; the seaction to penetration of hookworm larvae is somewhat more severe, often with the formation of papules or vesicles; it may last for a couple of weeks or longer if there is a secondary infection. The cutaneous larva migrans reaction caused by larvae of Ancylostoma braziliense is characterized by a reddened papule at the site of entry; the larva forms a reddened serpiginous tunnel, at first covered with vesicles, later dry and crusted, which advances at the rate of a few millimeters to centimeters a day. The area itches intensely; without treatment infection may persists for several weeks or months. Strongyloides larvae may produce similar cutaneous lesions, but because of their more rapid subcutaneous movement, they are referred to as "larva currens."

Presence of adult Loa loa beneath the skin may be indicated by a thin raised reddened line, a few centimeters in length. The adult female Dracunculus also may be visible beneath the skin but usually produces no reactior until about to larviposit, when a vesicle forms over the point at which the worm is about to break through the skin. In onchocerciasis, the presence of microfilariae in the skin sometimes elicits an acute pruritic inflammatory reaction resembling erysipelas and usually confined to the face, neck and ears; repeated acute attacks may result in n chronic lichenification, with hyperpigmentation and fissuring.

The migration of Sarcoptes scabiei through the skin produces lesions resembling those of cutaneous larva migrans, but frequently seen in parts of the body that have no contact with soil. This organisms is highly infectious, and localized hospital epidemics have been reported. The mites invade the upper layers of the epidermis, in which they form sinuous burrows. They seem to have a predilection for the interdigital, popliteal and inframammary folds the and groin. Intense itching, with formation of small vesicles and crusting of the chronic excoriated lesions, is typical. The larvae of the horse botfly, Gasterophilus, produce similar cutaneous lesions in man. Pediculosis in hypersensitive persons, usually as a result of repeated exposure, may give rise to a severe localized reaction, with reddish papules at the feeding site, and surrounding vesiculation and a weeping dermatitis. Bronzing of the affected area may persist following healing. The chigoe flea, Tunga penetrans, produces local pruritus as it lies partly buried in the skin of the toes or elsewhere on the body; secondary infections by clostridia are not uncommon. See also Rash.

Diarrhea. Diarrhea in parasitic diseases may be of diverse etiologies. In kala-azar, infiltration of the submucosa with leishmania-containing macrophages may lead to mucosal ulceration and diarrhea. Plugging of the mucosal capillaries with parasitized red blood cells may lead, in falciparum malaria, to a watery diarrhea so profuse as to suggest cholera. Blood, containing parasitized red cells, may be found in the stools. The diarrhea or dysentery is usually accompanied by nausea and vomiting. Mucosal ulceration is amebiasis or balantidiasis may produce diarrhea or, if the ulceration is more extensive, dysentery (q.v). Isospora develops within the epithelial cells of the lower ileum and cecum. Infection is self-limited, lasting usually a month or less, and often asymptomatic. In some cases there is mild abdominal pain, nausea and vomiting, and diarrhea. Giardia infections may be asymptomatic, accompanied by a mild mucoid diarrhea, or, like Isospora, may give rise to a full-blown malabsorption syndrome with steatorrhea.

Development of the cysticercoids of Hymenolepis nana within the intestinal villi may elicit a mucous diarrhea. Maturation of Trichinella witnin the wall of the duodenum and jejunum produces nausea, vomiting, colicky abdominal pain and diarrhea, starting about 24 hours after infection and lasting up to five days. In Strongyloides infections there may be mild diarrhea, alternating with periods of constipation, or the diarrhea may be severe and prostrating. In heavy infections ulceration and sloughing of the intestinal mucosa may take place, with dysentery and often with secondary bacterial infection and fever. Capillaria philippinensis infections may result in profuse diarrhea, malabsorption and a protein-wasting enteropathy.

In schistosomiasis mansoni and japonicum, there is diarrhea, presumably of toxic origin, associated with nausea, vomiting, hepatic tenderness, fever, eosinophilia and an urticarial rash, during the period while the worms are maturing in the liver sinusoids. Somewhat later, with the beginning of egg deposition in the intestinal wall, there may be a profuse diarrhea or dysentery.

In persons previously uninfected, the onset of a heavy hookworm infection may be marked by nausea, vomiting, epigastric or midabdominal tenderness, and diarrhea. The diarrhea is presumably caused by toxicity or hypersensitivity, although mechanical irritation may play some part. The same may be said of the diarrhea in heavy whipworm infections. In fasciolopsiasis, diarrhea, which usually has its onset about a month after infection takes place, is characterized by the passage of stools containing much undigested food; severe infections are accompanied by symptoms of severe malnutrition, with edema of the face, abdominal wall and lower extremities, ascites and prostration. The diarrhea sometimes seen in Taenia and broad fish tapeworm infections and in infections with the smaller intestinal flukes may be related to local irritation.

Dysentery. Acute amebic dysentery is characterized by the passage of six to eight or sometimes a dozen or more mucoid blood-flecked stools a day. There may be generalized abdominal pain and tenderness if the entire colon is involved, tenderness over McBurney's point, nausea and vomiting with cecal infection, or tenesmus, with relief of the accompanying pain after evacuation if the rectosigmoid is the main diseased area. An untreated attack lasts a few days to several weeks and usually subsides spontaneously to recur after an interval of some days to several years. Between attacks the patient may be constipated. Balantidial dysentery is similar to amebic, and as in the latter disease, many infections are asymptomatic. Dysentery accompanying kala-azar, falciparum malaria infections, strongyloidiasis and schistosomiasis is mentioned in the discussion of Diarrhea.

Edema. Circumorbital edema, possibly resulting from vasculitis provoked by the migrating larvae, is frequently seen in the early stages of trichinosis; there also may be edema of the hands. Unilateral circumorbital edema, with local pruritus and sometimes intense pain, results from passage of the adult Loa loa across the eyeball or lid. Passage of the worm across the eyeball takes from less than half a minute to as long as 10 minutes; the resulting inflammatory changes usually persist for several days. Calabar swellings (q.v.) are also seen in loiasis, while localized edema of the face, neck, ears, etc., accompanied by intense pruritus and erythema, may be recurrent in onchocerciasis. Ocular sparganosis is not uncommon in some areas as the result of poulticing eye lesions with split raw infected frogs. In the subcutaneous tissues around the eye, the sparganum produces a violent tissue reaction with edema; retrobulbar development of the sparganum may cause protrusion of the eyeball and consequent corneal ulceration. Areas of localized hard edema, of uncertain etiology, are frequently seen in Chagas' disease, occurring after appearance of the chagoma (q.v.). The most common type is unilateral edema of the eyelids (Romana's sign) (q.v.). Edematous patches may develop elsewhere on the body, especially involving the abdominal wall, pupic area, scrotum and legs. Of equally obscure origin is the edema of the hips, legs, hands and face that many accompany the acute stage of African sleeping sickness. Edema of the face and legs, with a pratuberant abdomen, is seen in severe hookworm infections, fasciolopsiasis and diphyllobothriasis, and may be related to malnutrition.

Elephantiasis. Filarial elephantiasis is a chronic enlargement of a limb, the scrotum, a breast or the vulva, with hyperplasia of the connective tissue and skin, a woody non-pitting edema, and thickened, coarsened skin, often with verrucous changes. In Malayan filariasis, elephantiasis is less severe and generally is confined to the lower limbs. Elephantiasis of the external genitalia in both sexes, and hypertrophy of the femoral lymph nodes, producing a peculiar conditions known as "hanging groin," have been reported from some areas in Africa where bancroftian filariasis is unknown, and are ascribed to onchocercal infection. In schistosomiasis haematobium, extensive egg deposition may lead to fibrosis, which blocks the lymphatic drainage, and to elephantiasis of the penis.

Eosinophilia. Eosinophilia is a consistent finding in helminth infections, though it may be quite variable in degree. In general, tissue parasites provoke a higher eosinophilia than those that live only in the lumen in the bowel.

As many physicians equate eosinophilia with parasitic disease, it may not be amiss to give the result of a Mayo Clinic study of 418 patients with an eosinophilia of 20 per cent or greater, as quoted by Harris (1979):

Diseases with Eosinophilia of Greater than 20 Per cent (Mayo Clinic 1944)

Disease	*Per cent*
Atopic diseases (vasomotor rhinitis, asthma, hay fever)	28
Lymphoproliferative diseases (lymphoma)	20
Dermatoses (pemphigus, dermatitis herpetiformis)	10
Nonparasitic infections (principally streptococcal)	11
Periarteritis nodosa (with lung involvement)	10
Parasitic infections	4
Nonlymphatic malignant tumors and leukemia	4
Miscellaneous	13

A marked eosinophilia (20 to 70 per cent or higher) is most frequently seen in trichinosis, strongyloidiasis, hookworm infection, visceral larva migrans, filariasis, schistosomiasis and fasciolopsiasis. Moderate eosinophilia (6 to 20 per cent) often accompanies trichuriasis, ascariasis, paragonimiasis, taeniasis and eosinophilic meningitis. It must be realized that eosinophilia is an index of host reaction to the parasite therefore, it will vary considerably from one patient to another.

Eosinophilia is not characteristic of any of the protozoan infections.

Tropical eosinophilia or eosinophilic lung disease is characterized by symptoms of chronic bronchitis or asthma, a marked eosinophilia, elevated erythrocyte sedimentation rate, paroxysmal cough or wheezing, malaise, easy fatiguability, anorexia and weight loss. The chest film may show diffuse patchy motling and transverse branching striations, most prominent in the midlung and basal lung fields, with enlargement of the hilar shadows. Occasionally there may be unilateral densities in the upper lung field which are suggestive of the picture seen in pulmonary tuberculosis. The disease is reported from such areas as India, Pakistan, Sri Lanka, China, Burma, Thailand, the Philippines, Malaysia and Indonesia, tropical Africa and the West Indies. Its etiology may be diverse. Filarial infection, with human or non-human species, is considered to be a common cause of this condition. Other agents may be the pulmonary stages of Ascaris, Strongyloides, Toxocara or other helminths. Treatment with conventional doses of diethyl-carbamazine is often effective; antimonial (stibophen) or arsenical drugs (neoarsphenamine) are sometimes effective when there is no response to diethylcar-bamazine.

Epididymitis. An early complication of filarial infection, often associated with orchitis, and with or without accompanying lymphangitis and fever.

Fever. Patterns of fever may be characteristic in malaria and kala-azar. but it is a mistake to suppose that they must conform to the "textbook" pattern.. A quotidian fever is often seen in the initial attack of vivax malaria, with two or more broods of parasites completing their exoerythrocytic cycle at different times, so that there may be a daily fever peak corresponding to rupture of the infected red cells and liberation of merozoites. Daily fever peaks usually are seen only for a few days; apparently within this time all broods of parasites become synchronized, and thereafter the fever cycle will exhibit a tertian periodicity. Quartan and falciparum malaria may likewise exhibit a quotidian or irregular periodicity during the first few days of the primary attack. Tertian fever is characteristic of vivax and ovale malaria. The paroxysm has an abrupt onset, usually initiated with a chill which varies from a moderate sensation of cold to the intense "bed-shaking" chill usually thought typical of malaria. The chill lasts up to an hour and the fever (to 104 or 105°) for two hours or longer, followed by a profuse sweat, during the course of which the temperature falls to normal over a period of an hour or so. The sweating stage usually is followed by sleep, and when the patient awakens he usually feels well. The next paroxysm is initiated approximately 48 hours after the onset of the previous one. Subtertian fever, seen in falciparum malaria, is so called because the cycle may more nearly approach 36 than 48 hours. There is usually no frank chill, and the febrile stage is prolonged, though the fever is not usually so high as in vivax; it may not fall to normal even in the intervals between paroxysms. There may be double peaks of fever in each 24-hour period, resembling the fever curve in kala-azar. There is often no well-defined sweating stage, though sweating may be continuous, periodic or completely absent. Quartan fever is seen in malaria caused by Plasmodium malariae. There is often a regular periodicity from the start; the paroxysms recur at 72-hour intervals; while similar to those of vivax malaria, they are generally more severe. The hot stage often lasts several hours and is frequently accompanied by nausea and vomiting; the sweating stage may be followed by prostration. Hyperpyrexia may develop as part of an attack of cerebral malaria or in the course of an apperently uncomplicated attack of falciparum malaria; as the result of injury to the heat-control center in the hypothalamus there is a rapid rise in temperature to 107° or higher, and death quickly ensues.

A doubly remittent or "dromedary" fever is frequently found in kala-azar. Febrile attacks, which last a few days to several weeks, are separated by afebrile periods of equal irregularity. At some time during the course of a febrile attack, there will usually be one or more days during which a double or triple rise to a temperature of 103 to 105° can be demonstrated during a 24-hour period.

An irregularly spiking fever with hepatic tenderness, suggestive of cholangitis, may be seen in amebic hepatitis, fascioliasis and acute Opisthorchis infections. The initial period of schistosome infection is likewise marked by irregular fever and hepatic tenderness, with nausea and vomitting, diarrhea and giant hives. An irregular fever, usually with evening peaks and night sweats, is an early finding in African trypanosomiasis, and a high remittent fever, lasting for several weeks, occurs early in Chagas' disease. A remittent fever, with temperature to 104 or 105°, frequently marks the stage of larval migration in trichinosis. Filarial fever may occur very early in the course of a filarial infection. There is usually a sudden onset, with fever ranging in the neighbourhood of 102 to 104° and remaining elevated for several hours to a couple of days, and gradually subsiding in the next several days. Attacks of lymphangitis and lymphadenitis (q.v.) usually accompany the febrile episodes.

Funiculitis. Inflammation of the spermatic cord is frequently an early symptom of filariasis.

Hematuria. In Schistosoma haematobium infections, beginning as soon as three months after infection or sometimes not until several years later, there may be intermittent hematuria. There is no dysuria, but there may be some frequency and also bladder pain following urination. Hematuria is often referred to as terminal hematuria, being limited to the last few drops of urine, blood being forced out as the bladder wall contracts.

Hemoglobinuria. Blackwater fever usually is seen in conjunction with an attack of falciparum malaria, generally in patients who have had previous attacks of malaria. The passage of reddish or red-brown urine signals a bout of intravascular hemolysis, which may occur once or repeatedly and may lead to severe renal tubular damage and anuria. The cause of blackwater fever is unknown; hypotheses include quinine sensitivity, glucose-6-phnsphate dehodrogenase deficiency in persons treated with primaquine and related drugs, and autohemolysis on the basis of antibodies formed against altered infected red cells.

Hepatitis. Amebic hepatitis has already been described under the heading of Abscess, Amebic.

Hepatomegaly. Any parasitic infection involving the liver may result in enlargement of that organ. Thus, amebic hepatitis or liver abscess, visceral larva migrans, liner-fluke infections and early schistosomiasis are all characterized by an enlarged and tender lives, which is also seen in some cases of falciparum malaria and acute neonatal toxoplasmosis as well as in kala-azar and Chagas' disease. Hydatid infections of the liver and Schistosoma mansoni and japonicum infections may result in an enlarged but usually non-tender liver. The hepatic fibrosis characteristic of chronic schistosome infections produces a clinical picture similar to that of Laennec's cirrhosis, often with splenomegaly.

Those helminth infections involving the liver are characterized also by an eosinophilia. In visceral larva migrans, a leukocytosis of up to 80,000 may be present, with a striking eosinophilia. In amebic abscess of the liver, on the other hand, though the white blood count may be in the range of 25,000, there is no eosinophilia.

Hives. Giant hives and other allergic symptoms, such as bronchial asthma, are commonly seen in ascariasis, and hives often appear during the first few weeks of schistosome infections.

Hydatid Thrill. In large unilocular echinococcus cysts of the abdominal viscera that are situated close to the abdominal wall, a characteristic thrill may be elicited by quick palpation or percussion.

Hydrocele. This is a common finding in areas where filariasis is endemic, developing as a sequel to repeated attacks of orchitis. If lymphatic varices develop in the cord and rupture into the scrotal sac, a condition known as lymphocele results.

Hydrocephalus. Although not so intimately associated with congenital toxoplasmosis as are chorioretinitis and cerebral calcifications, hydrocephalus or microcephaly is commonly seen in this condition.

Hyperpigmentation. Kala-azar derives its name from intensification of the pigmentation of the skin over the cheeks and temples, and around the mouth. It is most obvious in dark-skinned races. In onchocerciasis, repeated attacks of allergic dermatitis may result in hyperpigmentation of the area, usually on the face, neck or ears. Bronzing and induration of the affected skin areas may oceur in chronic pediculosis (vagabond's disease).

Jaundice. Obstructive jaundice may be seen in severe liver fluke infections but is not characteristic of light infections or those of moderate intensity. The symptoms of the falciparum malaria syndrome known as bilious remittent fever include acute epigastric pain, nausea and vomiting, marked enlargement and tenderness of the liver, with jaundice appearing on about the second day. Diarrhea, a high remittent fever and oliguria are usually seen, and death may result from renal or hepatic failure.

Kerandel's sign. Noted in the stage of central nervous system involvement in African sleeping sickness, this sign may be elicited by pressure on the palm of the hand or over the ulnar nerve and consists of severe pain which occurs shortly after the pressure has been relieved.

Leukocytosis. Leukocytosis seldom continues throughout the course of any of the parasitic infections. In amebic hepatitis or abscess there may be a white blood cell count of 25,000 to 30,000, with 70 to 80 per cent polymorphonuclear neutrophils. In visceral larva migrans a leukocytosis of up to 80,000 has been reported, with an eosinophilia of 20 to 80 per cent. Trichinosis may be characterized by a white count of 30,000 early in the infection, and strongyloidiasis by one nearly as elevated; these usually decline and may be followed by leukopenia. Leukocytosis early in the course of infection, followed by leukopenia with a relative monocytosis. is common to many protozoan and helminth infections.

Leukopenia. A white cell count of 4000 or less, with a relative monocytosis, generally is seen throughout the course of kala-azar, sometimes terminating in agranulocytosis.

In malaria, a leukopenia of 3000 to 6000, with a relative monocytosis, characterizes the afebrile periods, while there may be a leukocytosis during the paroxysm.

Lymphadenitis. In filariasis the femoral and epitrochlear nodes are most commonly involved, as are the axillary and inguinals. The nodes are enlarged, painful and tender during an acute attack of lymphangities and tend to remain enlarged between attacks. A condition resembling infectious mononucleosis sometimes is seen in the acute stage of toxoplasmosis, with fever, weakness, malaise, generalized adenopathy and sometimes a rash. There may be generalized lymphadenitis without fever or other symptoms. In the early stages of African trypanosomiasis there may be generalized adenopathy; the glands of the posterior cervical triangle are most conspicuously affected (Winterbottom's sign). Generalized adenopathy is seen in the acute stage of Chagas' disease.

Lymphangitis. Acute lymphangitis is an early symptom of filarial infection. It usually is accompanied by fever and may affect the limbs, breast or scrotum. When it occurs on a limb, it is usually centrifugal in development, starting at a lymph node and progressing distally. The course of the lymphatic is readily seen because of local distention and erythema. Centrifugal spread of the lymphangitis is the reverse of that seen in bacterial lymphangitis (blood poisoning), in which the infection extends proximally from the point of origin.

Lymphocytosis. A relative or absolute lymphocytosis, unusual in parasitic infections, is, however, generally seen in Chagas' disease. Initially there may be a slight leukocytosis, usually followed by leukopenia.

Lymph varices. Dilatations of the lymphatic vessels may occur secondarily to lymphatic blockage in filariasis. They are most frequently seen in the inguinal and femoral areas, or other lymphatic tracts may be affected. The soft lobulated swellings may rupture and drain. When this occurs on the scrotum, a chronic condition known as lymph scrotum may develop.

Melena. Upper gastrointestinal bleeding of a degree sufficient to produce melena is rare in parasitic disease; it is mentioned here primarily with strongyloidiasis in mind. Infection with Strongyloides stercoralis may run the gamut from a complete lack of symptoms to those infections of the duodenum and jejunum which are so extensive as to produce ulceration of the mucosa. with blood loss to the point of clinical anemia, and with melena. Especially in immunosuppressed patients, the coincidental finding of anemia with melena and a significant eosinophilia should stimulate a thorough search for this sometimes elusive parasite.

Meningoencephalitis. Invasion of the central nervous system by trypanosomes is characterized in African sleeping sickness by increasing symptoms of meningoencephalitis. There may be quite variable sensory and motor changes, personality disorders, headache, confusion, drowsiness and finally coma. Similar but milder symptoms are seen in Chagas' disease. Minor neurologic symptoms are seen during almost any attack of malaria (*i.e.*, headache. disorientation), and cerebral malaria may develop as a complication of any type of malaria. It is characteristic of falciparum malaria, however, and may develop slowly with increasing headache and drowsiness over several days or may present as a sudden coma or other acute mental disturbance. There may be signs of meningeal irritation; symptomatology is quite varied, depending upon the brain areas affected. If the cord is affected, the symptoms may be suggestive of multiple sclerosis.

Amebic meningoencephalitis, caused by invasion of the central nervous system by ordinarily free-living ameboflagellates of the genus Naegleria and possibly other amebae, is an acute, rapidly progressive infection, apparently acquired while swimming or diving in fresh-water lakes or pools. It is characterized by fever, headache, mental confusion and coma, and death frequently occurs within a few days of onset.

Eosinophilic meningoencephalitis, seen in various parts of the Pacific area in recent years, is believed, on strong epidemiologic grounds, to be symptomatic of infection with Angiostrongylus cantonensis and thus a form of larva migrans infection. It is characterized by fever, headache, stiff neck, and increased cells (mainly eosinophils) in the spinal fluid. It is generally a mild and self-limited infection.

Microcephalus See *Hyderocephalus*

Monocytosis. A relative or absolute monocytosis is a frequent finding in both protozoal and helminthic infections.

Myocarditis. Myocardial infection is characteristic of Chagas' disease and is seen in about 50 per cent of chronic cases. Cardiac failure may come on slowly, although in infants it tends to occur in the early acute stage. There may be pericardial effusion. Congestive heart failure also has been reported in African trypanosomiasis the probably in the Rhodesian form of the disease. In African typanosomiasis the etiology of the heart failure is not as apparent. Myocarditis, occasionally severe enough to cause death, has been reported. in trichinosis. It is the result of migration of the larvae through the myocardium, in which they do not encyst. Myocarditis also may be seen in acute toxoplasmosis in both infants and adults, the result of invasion of the myocardium.

Myositis. Although myositis is a nonspecific symptom of many febrile illnesses, severe myositis is characteristic of the stage of larval migration in trichinosts. If accompanied by circumorbital edema, eosinophilia and a history of consumption of improperly cooked pork, the diagnosis may be made with some certainty. Myositis, usually involving a single muscle group, may also occur in Sarcocystis infection.

Neurologic symptoms. Neurologic symptoms in trypanosomiasis, malaria and amebic and eosinophilic meningoencephalitis are discussed under Meningoencephalitis. Variable neurologic symptoms may occur in schistosomiasis when eggs carried by the bloodstream to the central nervous system lodge there and provoke a granulomatous reaction. Neurologic and other symptoms caused by embolization of eggs are more common in Schistosoma japonicum infection than in the other two species, while in S. mansoni and S. haematobium, eggs are found more frequently in the spinal cord than in the brain, perhaps because of ectopic wanderings of the adult worms. In S. japonicum infection, there may be severe neurologic symptoms, including coma and paresis, during the incubation period or first few weeks after infection. Transitory neurologic symptoms of a variable nature may be caused by the migration of ascarid and trichina larvae in the central nervous system; hemiplegia and focal epileptic attacks have been reported in trichinosis. Hydatid, coenurus and cysticercus cysts may develop within the central nervous system, where they may produce symptoms related to a space-occupying lesion or, if within the ventricular system, internal hydrocephalus. Cysticercus larvae may give rise to epileptiform seizures, as may Sparganum proliferum and adults of Paragonimus westermani that have gone astray.

Nodules, subcutaneous. Lipoma-like subcutaneous nodules include onchocercomas (q.v.) and cysticercus and coenurus larvae. The cysticercus larva of Taenia solium develops most frequently in the subcutaneous tissues, where it forms nodules 0.5 to 3.0 centimeters in diameter. In almost half the recorded human cases of coenurus infection, the larvae have been found in the subcutaneous tissues; others have been recorded from the brain, spinal cord and eye. Echinococcus cysts also may be found in the subcutaneous tissues. Spargana also form subcutaneous nodules, somewhat elongate and several centimeters in length, which may resemble lipomas, but they may move through the subcutaneous tissues at irregular intervals and often cause pain. Sparganum proliferum may develop as branched or multiple nodules and invade the viscera. (See also Edema for a discussion of ocular sparganosis). Larvae of the botfly Hypoderma migrate through the subcutaneous tissues, finally coming to rest beneath the skin, where they produce elongate nodules several centimeters in length. Considerable pain may accompany migration, but the resting nodule is seldom painful or pruritic. The human botfly, Dermatobia hominis, burrows into the skin and subcutaneous tissues, producing an intensely pruritic papular lesion, which has the appearance of a furuncle. There is a small central opening, from which comes a serous exudate and through which the posterior end of the larva may protrude from time to time. Secondary infection is common.

Obstruction, intestinal. Ascaris, especially in children, may produce complete intestinal obstruction, with accompanying abdominal pain, vomiting, distention and hyperperistalsis. Partial or complete intestinal obstruction may also characterize infection by Angiostrongylus costaricensis.

Ocular Symptoms. See *Conjunctivitis, Edema and Visual Difficulties.*

Onchocercoma. Adult worms of Onchocerca voleulus lie in coiled masses beneath the skin, completely enclosed in a fibrous tissue capsule. They are from a few millimeters to several centimeters in diameter, generally freely movable, and resemble lipomas. In Mexico and Guatemala they frequently occur beneath the patient's scalp; in Africa most of them occur on other parts of the patient's body.

Orchitis. Filarial orchitis may occur early in the disease and at times in the absence of lymphangitis or fever; repeated attacks lead to hydrocele.

Pain. Abdominal pain, generally vague or ill-defined, is said to accompany many of the intestinal parasitic infections. The presence or absence of such tenuous pains is of no value from a diagnostic standpoint. Epigastric pain, sometimes with nausea and vomiting, may be seen in giardiasis, trichinosis and strongyloidiasis; it is related to the duodenitis and jejunitis provoked by these infections. Moderate to severe abdominal pain is seen in acute amebic colitis; it may be confined to the cecal area or may be generalized. Angiostrongylus costaricensis may give rise to similar symptoms. In ascariasis, severe pain may signal intestinal obstruction (q.v.), perforation and peritonitis (q.v) or bile duct blockage.

Muscle pain in trichinosis is discussed under Myositis, and the delayed pain sensation seen in African trypanosomiasis under the heading of Kerandel's Sign.

Peritonitis. Penetration of Ascaris through the wall of the intestine usually leads to generalized peritonitis, with pain, marked distention, generalized abdominal tenderness, and free air under the diaphragm, detectable by X-ray. In severe amebic dysentery, ulcers may erode through the wall of the intestine and cause peritonitis.

Pneumonitis. Pneumonitis is characteristic of severe Ascaris infection and is caused when the worm larvae break out of the capilliaries into the alveoli, whence they are coughed up to be swallowed, beginning the intestinal phase of the disease. Symptoms and signs are first noted one to five days after the eggs are ingested and consist of cough, fever, respiratory distress and the physical and X-ray signs of a bronchopneumonia; in severe cases there may be complete consolidation of one or more lobes. The pneumonitis usually clears within a week or two; it may be accompanied by high eosinophilia and an urticarial rash. Similar signs and symptoms may accompany the corresponding stage in strongyloides infection, although the pneumonitis is generally not so severe. In hookworm infection there is seldom a clear-cut pneumonitis at this stage, but a cough is frequently present. In schistosomiasis there may likewise be a transitory cough, sometimes with hemoptysis and frequently with dyspnea, during the stage of migration through the lungs.

Pneumocystis carinii causes an interstitial plasma cell pneumonia, mainly in newborn and young children. Some cases have been reported in adults. The X-ray picture is that of bronchopneumonia. Pneumonitis may be a part of acute toxoplasmosis in infants during the neonatal period, and more rarely in adults. Atelectatic pneumonitis may be seen in amebic abscess, the result of pressure from the elevated right diaghragm the abscess may erode through the diaphragm to produce a right lower lobe infection, infrequently an abscess may develop primarily in the lung. If there is erosion into a bronchus, there may be expectoration of abscess material, a light reddishbrown, or "anchovy paste," colour.

Proteiunria. In falciparum malaria, proteinuria, with hyaline and granular casts in the urine, is common. Rarely there may be oliguria or anuria, usually accompanying an attack of blackwater fever (q.v.). The nephrotic syndrome, with proteinuria, is sometimes seen in quartan malaria.

Pruritus ani. The nocturnal pruritus that accompanies pinworm infection varies considerably in degree. probably depending upon hypersensitivity of the host. In some persons there is no noticeable itching, whereas in others it may be sufficiently severe to interfere with rest. Anal pruritus may be associated with active migration of gravid proglottids of Taenia saginata out of the anus.

Pulmonary symptoms, chronic. In all three types of schistosomiasis, but especially in S. haematobium infections, ova may be carried to the lungs, where pseudotubercle formation around them may produce a radiological picture suggestive of miliary tuberculosis, and increasing fibrosis may lead to cor pulmonale. Paragonimiasis is characterized by chronic cough, the production of thick blood-specked sputum or sometimes frank hemoptysis, and increasing dyspnea. X-rays may show patchy infiltrates, rounded shadows suggestive of coin lesions, calcifications and pleural thickening or effusion. In pulmonary echinococcosis, cough is usually the first symptom. There may be increasing dyspnea; with erosion of blood vessels there will be hemoptysis, and with obstruction there results secondary bacterial infection and fever. If the cyst ruptures into a bronchus, the contents may be coughed up, or the patient becomes asphyxiated.

Rash. An allergic urticarial rash or hives (q.v.) is often seen in the early stages of schistosome infection and in ascariasis. A macular or maculopapular eruption may occur early in the course of a trichina infection, and one of the variants of acute toxoplasmosis is a typhus-like fever, with a macular rash, prostration, and sometimes stupor and cardiac decompensation. During attacks of fever in the early stages of African trypanosomiasis there may be an irregular blotchy rash, often annular in appearance. The individual patches may be several inches across; they tend to fade in a few hours, and reappear at irregular intervals.

Retinochoroiditis See *Chorloretinitis*

Romana's sign. Unilateral palpebral edema, involving both upper and lower eyelids, appears only in the course of an infection with T. cruzi. The edema is hard and non-pitting; it may remain confined to the eyelids or may spread down to involve the cheek and neck. It may subside promptly or persist for weeks or month.

Shock. When shock complicates falciparum malaria, the

patient is pale, with a cold and clammy skin, thin fast pulse, and low blood pressure. There is often acute abdominal pain, vomiting and diarrhea. The etiology may be one of primary adrenal failure, through parasite-induced ischemia or infarction or it may be secondary to reduced blood volume and blood pressure caused by widespread vascular injury. Rupture of an echinococcus cyst may lead to anaphylatic shock.

Splenomegaly. As part of the generalized lymphoid hyperplasia in both African and American trypanosomiasis, splenomegaly may be observed. In kala-azar the spleen is said to enlarge downward about an inch per month, and it may extend into the pelvis. It is non-tender and reverts to normal size after effective therapy. The spleen enlarges during an acute attack of malaria and is usually palpable within two weeks after onset. Between attacks it may regress in size, and in adults it may become fibrotic and smaller than normal. In children, repeated attacks may lead to great enlargement of the organ, which may reach the pelvis. The "splenic index" as a guide to endemicity of malaria, obtained by examination of a population for evidence of enlarged spleens, obviously must be derived only through examination of children. The spleen is usually tender during an acute attack of malaria, and tenderness may be apparent before the organ can be palpated. Splenic infraction or rupture occurs rarely. In Schistosoma mansoni and S. japonicum infections, splenomegaly is secondary to hepatic fibrosis brought about by egg deposition in the liver, and portal hypertension.

Splinter hemorrhages. Sometimes occurring during the stage of active larval migration in trichinosis, these hemorrhages are a sign of vascular injury.

Steatorrhea. Malabsorption, characterized by the presence of fat in the stool, is seen in certain parasitic infections. The most common of these is giardiasis, which may make its presence known by flatulence and the production of foul-smelling fatty stools. The less common and self-limited Isospora belli infections are also characterized by steatorrhea, as are those caused by Capillaria philippinensis.

Tachycardia. A fast pulse is noted early in both African and American trypanosomiasis. In Chagas' disease it persists into the subacute and chronic stages, where it may be associated with heart block, Stokes-Adams syndroma and fibrillation.

Ulcers, cutaneous. In leishmanial and trypanosomal diseases there is a primary multiplication at the site of infection. In Leishmania tropica-complex infections there is first a papule at the site of infection, which gradually transforms into a shallow ulcer with raised edges. The Chiclero ulcer of Southern Mexico and Central America is similar to oriental sore, except when it occurs on the eat, where it may erode the pinna. L. braziliensis first produces cutaneous ulcerations, which may, through extension or metastasis, come to involve the nasal mucosa, the soft and the hard palate, the nasal septum, the pharynx and the larynx. In blacks, granulomatous rather than ulcerative lesions are generally seen. In African sleeping sickness there may be a firm tender raised lesion, up to 3 cm. or more in diameter, at the site of infection. This "trypanosomal chancre" is painful or pruritic, but like the chagoma (q.v.) it does not apparently ulcerate unless secondarily infected. Ulcerative cutaneous lesions rarely are seen in amebiasis, either in the perianal region or in the skin surrounding fistulas or surgical drainage incisions from hepatic obscesses. A rounded ulcer, 2 mm. to several centimeters in diameter, marks the place at which the guinea worm discharge its larvae. In the centre of the ulcer, a portion of the worm may be visible. There is often secondary infection, and a painful localized reaction may persist until discharge of the larvae is complete.

Urethritis. Trichomonas vaginalis has been found in up to one third of cases of "nonspecific" urethritis in the male.

Vaginitis. A prolific, irritating, green or yellowish, thin discharge is seen in Trichomonas vaginalis infection; the vagina may be diffusely congested, or punctuate hemorrhagic spots may be seen. The organisms may be present in asymptomatic individuals. Pinworms may migrate from the angus and enter the vagina, where they produce a temporary, intense pruritus in some children.

Visual difficulties. Associated with parasitic disease include circumorbital edema (q.v.), conjunctivitis (q.v.) and chorioretinitis (q.v.). Ascarid larvae (both those of Ascaris lumbricoides and Toxocara) may invade the eye, producing iritis or other symptoms. Patients infected with Onchocerca actually may be aware of the intraocular movement of the microfilariae, and lesions of the anterior chamber, iris, ciliary body, choroid, and retina, developing in this condition, may lead to diminution of vision or total blindness. Cysticercus and coenurus larvae may develop within the eye. Ophthalmomysiasis may occur.

Winterbottom's Sign See *Lymphadenitis*

X-ray evidence of parasitic disease. Amebiasis: Amebic granulomas of the large bowel simulate carcinoma in barium enema studies. Cecal amebiasis tends to produce a funnel-shaped deformity of that portion of the bowel as seen in barium enema studies. In amebic abscess of the liver there may be elevation of the right diaphragm and sometimes right lower lobe pneumonitis. Intravenous sodium diatrizoate (Hypaque) infusion and tomography of the liver may show the wall of the amebic abscess. Giardiasis: Evidence of intestinal malabsorption. Toxoplasmosis: Intracerebral calcifications. Pneumocystosis: Pneumonitis, typically sparing lateral margins of the bases. Dracunculiasis, filariasis and loiasis: Calcified worms may be seen in the tissues. Ascariasis: Pneumonitis. Adult worms may be seen as cylindrical empty spaces in the barium-filled bowel in a small bowel series. Strongyloidiasis and hookworm infection: Pneumonitis. Loss of mucosal markings and a tubular deformity of the duodenum and jejunum may be seen in small bowel studies on patients with strongyloidiasis. Cysticercosis: Calcified cysts in the subcutaneous tissues, muscles, brain. Echinococcosis: Well-defined. rounded masses may be seen in the lung parenchyma; sometimes a fluid level is visible within them. Hepatic cysts are visible only if calcification of the wall has taken place. Sometimes calcified daughter cysts are seen. (Hepatic photoscanning with the use of radioactive isotopes will reveal non-calcified cysts as well.) Hydatid cysts of bone produce extensive intramedullary erosion, demonstrable by X-ray. Paragonimiasis: Patchy infiltrates or rounded densities in the parenchyma, pleural thickening, or fluid. Schistosomiasis: Pulmonary fibrosis, or a picture suggestive of miliary tuberculosis. Corpulmonale with dilatation of the pulmonary artery and its main branches, right ventricular hypertrophy. S. haematobium infections: Calcification in the wall of the bladder, hydronephrosis, hydroureter.

REFERENCE

Harris, E.D. 1979, In: Case records of the Massachusetts General Hospital, N. Engl. J. Med, 301:256-263.